The WBF Book Series: Volume 2

Applying ISA-88 in Discrete and Continuous Manufacturing

The WBF Book Series: Volume 2

Applying ISA-88 in Discrete and Continuous Manufacturing

By WBF

Edited by
William Hawkins
Dennis Brandl
Walt Boyes

 MOMENTUM PRESS

Momentum Press, LLC, New York

Applying ISA-88 in Discrete and Continuous Manufacturing
Copyright © Momentum Press®, LLC, 2010

First published in 2010 by
Momentum Press®, LLC
222 East 46th Street, New York, N.Y. 10017
www.momentumpress.net

ISBN-13: 978-1-60650-200-6   (hard back, case bound)
ISBN-10: 1-60650-200-X   (hard back, case bound)

ISBN-13: 978-1-60650-202-0   (e-book)
ISBN-10: 1-60650-202-6   (e-book)

DOI forthcoming

Volume 2 in the WBF Book Series by WBF, edited by William Hawkins, Dennis Brandl, and Walt Boyes, published by Momentum Press®, LLC

Cover Design by Jonathan Pennell
Interior Design by Scribe, Inc. (www.scribenet.com)

First Edition: December 2010

10 9 8 7 6 5 4 3 2 1

Printed in Taiwan

Contents

Figures

Tables

WBF Foreword

The purpose of this series of books from WBF, The Organization for Production Technology, is to publish papers that were given at WBF conferences so that a wider audience may benefit from them.

The chapters in this series are based on projects that have used worldwide standards—especially ISA-88 and 95—to reduce product variability, increase production throughput, reduce operator errors, and simplify automation projects. In this series, you will find the best practices for design, implementation, and operation and the pitfalls to avoid. The chapters cover large and small projects in a wide variety of industries.

The chapters are a collection of many of the best papers presented at the North American and European WBF conferences. They are selected from hundreds of papers that have been presented since 2003. They contain information that is relevant to manufacturing companies that are trying to improve their productivity and remain competitive in the now highly competitive world markets. Companies that have applied these lessons have learned the value of training their technical staff in relevant ISA standards, and this series provides a valuable addition to that training.

The World Batch Forum was created in 1993 as a way to start the public education process for the ISA-88 batch control standard. The first forum was held in Phoenix, Arizona, in March of 1994. The next few years saw growth and the ability to support the annual conference sessions with sponsors and fees.

The real benefit of these conference sessions was the opportunity to network and talk about or around problems shared by others. Papers presented at the conferences were reviewed for original technical content and lack of commercialism. Members could not leave without learning something new, possibly from a field thought to be unrelated to their work. This series is the opportunity for anyone unable to attend the conferences to participate in the information-sharing network and learn from the experiences of others.

ISA-88 was finally published in 1995 as *ISA-88.01-1995 Batch Control Part 1: Models and Terminology*. That same year, partially due to discussions at the WBF conference, ISA chartered ISA-95 to counter the idea that business people should be able to give commands to manufacturing equipment. The concern was that business people had no training in the safe operation of the equipment, so boardroom control of a plant's fuel oil valve was really not a good idea. There were enough CEOs smitten with the idea of "lights-out" factories to make a firewall between business and manufacturing necessary. At the time, there was a gap between business computers and the computers that had infiltrated manufacturing control systems. There was no standard for communication, so ISA-95 set out to fill that need.

As ISA-95 began to firm up, interest in ISA-88 began to wane. Batch control vendors made large investments in designing control systems that incorporated the models, terminology, and practices set forth in ISA-88.01 and were ready to move on. ISA-95 had the attention of vendors and users at high levels (project-funding levels), so the World Batch Forum began de–emphasizing batch control and emphasizing manufacturing automation capabilities in general. This was the beginning of the transformation of WBF into "The Organization for Production Technology." Production technology includes batch control.

The WBF logo included the letters "WBF" on a map of the world, and since this well-known image was trademarked, the organization dropped the small words "World Batch Forum" entirely from the logo after the 2004 conference in Europe. WBF is no longer an acronym. Conferences continued annually until the economic crash of 2008. There was no conference in 2009 because many companies, including WBF, were conserving their resources.

WBF remained active and solvent despite the recession, so a successful conference was held in 2010 using facilities at the University of Texas in Austin. Several papers spoke of the need for procedural control for continuous and discrete processes. The formation of a new ISA standards committee (ISA106) to address this need was announced as well. Batch control is not normally associated with such processes, but ISA-88 has a large section on the design of procedural control. There is a need for a way to apply that knowledge to continuous and discrete processes, and some of those discussions will no doubt be held at WBF conferences, especially if the economy recovers. We would like to invite you to attend our conference and participate in those discussions.

WBF has always been an organization with an interest in production technologies beyond batch processing, even when it was officially "World Batch Forum." Over the years, as user interests changed, so has WBF. We have not lost our focus on batch; we have widened our view to include other related technologies such as procedural automation. We hope you will find these volumes useful and applicable to your needs, whatever type of process you have, and if you would like more information about WBF, we are only a simple click away at http://www.wbf.org.

William D. Wray, Chairman, WBF
Dennis L. Brandl, Program Chair, WBF
August 2010

Foreword by Walt Boyes

Many years ago, some dedicated visionaries realized that procedure-controlled automation would be able to codify and regularize the principles of batch processing. They set out on a journey that eventually arrived at the publication of the batch control standard ISA-88 and the development of the manufacturing language standard ISA-95.

Many end users have benefited from the work of these visionaries, who founded not only the ISA-88 Standard Committee but also the WBF. WBF has been an unsung hero in the conversion of manufacturing- to standards-based systems.

Today, WBF continues as the voice of procedure-controlled automation in the process and hybrid and batch processing industries. The chapters that make up this book series provide a clear indication of the power and knowledge of the members of WBF.

I have been proud to be associated with this group of visionaries for many years. *Control* magazine and ControlGlobal.com are and will continue to be supporters of WBF and its aims and activities.

I would like to invite you to come and participate in WBF, both online and at the WBF conferences in North America and Europe that are held annually. You will be glad you did. You can get more information at http://www.wbf.org.

Walt Boyes, ISA Fellow
Editor in Chief
Control magazine and ControlGlobal.com

Preface

This book is primarily about ISA-88.01 as it applies to non-batch control. The ISA Standards and Practices Board was asked to charter a batch control models and terminology group by Tom Fisher in 1988, and so the committee was called ISA88. I was active in that group from about 1990 to 1996. That was the golden age of standards, before mutual fund growth caused many in upper management to focus on growing the stock value via the bottom line and to lose track of what it took to maintain a manufacturing company. In other words, people couldn't get time for standards work or money for travel. It only got worse as time went on.

As a batch group with vendors, users, and consultants from that industry, we focused on batch processes. Only later did people begin to notice that we had created a pretty good set of engineering practices for *any* kind of automation of procedural control. We had tried to maintain a view of manual procedures during the deliberations of ISA88, but most people saw it as aimed at automation.

Many of the chapters in this book were written during the transition from ISA-88 as a batch standard to a procedural control standard. That name "batch" wouldn't let people see outside the box they'd put themselves in with a batch mindset. There is hope. We await the deliberations and results of ISA106, which has been chartered to address the needs of continuous and discrete processes.

It is ironic that process control, as it went through the pneumatic, electrical, and computerized stages, aimed first at the refineries. Once the refineries were happy, the vendor's growth opportunity was batch control. In the seventies, a DCS could be had with a Batch Package. Dr. Richard Lasher of Exxon called these packages "a FORTRAN compiler and a cheery 'good luck.'" ISA-88 brought a new order to batch control, causing users to demand better products and vendors to respond. Now that batch control systems make their customers happy (for the most part), vendors are looking at selling procedural control to the continuous and discrete markets.

ISA-88 will not help control system sales beyond Batch until vendors stop talking about Batch Engines, Managers, and Historians and replace Batch with Procedural Control (see Chapter 17).

Dave Chappell from P&G asked ISA to charter part 5 of ISA-88 for packaging machinery, which is a logical extension to batch control because many batch products require packaging to be sold. He also stirred up the Organization for Machine Automation and Control (OMAC), who created PackML and Make2Pack. Some people in these organizations, notably Adam Maki, found great value in ISA-88's exception handling state machine. They wanted to add some states (e.g., dividing Running into Producing and Standby) and needed to find a standards group that would let them do that. It really does the user no good to let each vendor come up with their own set of states and behavior. Think of the training issues.

Please read Dave Chappell's introduction to this book, which directly follows this Preface.

Chapters 1, 10, 12, and 14 address PackML and the needs of packaging machines. Chapter 10 by Ian McDonald of Unilever adds a business perspective to the discussion by one of the best in the business.

Chapters 2 and 3 discuss non-stop batch processing, as invented by Dennis Brandl. The process considers individually packaged foods as they move from one unit to the next, doing something different to the product in each stationary unit. Chapters 4, 6, 7, and 8 predict the effects of taking ISA-88.01 to the "next level," which means above batch to continuous and discrete processes. As a great fan of batch control, I consider continuous and discrete to be special cases of batch, and so they are a level down from batch control.

Chapter 5 is not concerned with the next level but with the construction of pharmaceutical plants and the extra needs of validation for federal regulations. It is well worth reading, whether your industry is regulated or not.

Chapter 9 discusses process information handling at the next level.

Chapter 11 covers statecharts for procedural control in exquisite detail. Willie Lotz is an isolated genius in my humble opinion, and so his work is not well known. His statecharts are definitive.

Chapter 13 adds network security and ISA-99 to the discussion. It is well worth reading if security is new to you.

Chapter 15 emphasizes compliance in a regulated industry for packaging.

Chapter 16 talks about the benefits of software modularity and introduces the concept of a Technology Evangelist.

I wrote chapter 17 to focus on the procedural control and exception handling that are needed for continuous and discrete control.

Bill Hawkins
Minneapolis, August 2010

Introduction

ISA-88 Concepts Applied to Manufacturing Techniques Other Than Batch

During the development of the ISA-88 batch standard in the late eighties, some of us on the committee evaluated the application of the batch concepts in this standard to determine whether or not they could effectively be applied to continuous and discrete manufacturing. The consensus was that the concepts fit very well and could be of valuable assistance to these other industries. Tom Fisher agreed but pointed out that if we tried to expand our charter beyond batch, the probability of us ever getting finished was greatly lessoned. Tom sure was right! The projected 2-year effort actually spanned 13 years to deliver ISA-88 parts 1 and 2 to satisfy the original charter. Good thing we stuck with batch, as I can't imagine where we'd be now if we had expanded our charter!

A few of us working in industries that used non-batch manufacturing did attempt to apply the ISA-88 concepts there. The results were quite interesting and taught many of us a lot about human nature and organizational power bases and momentum. To say the intrusion of this batch thinking was *not* well received in areas outside of batch would be a gross understatement. During the time of the ISA-88 development, batch was the most significant "pain" point for manufacturing. The other types of manufacturing were working just fine, thank you, so "don't mess with success" was the mantra of those organizations. Anyone who tried to evangelize the virtues of the ISA-88 batch concepts was quickly shown the door and was lucky if no bodily harm occurred! The current approaches for these manufacturing techniques were institutionalized, and the organizations that applied them were firmly in place and not open to new methods that might disrupt their comfortable lives. Understanding the ISA-88 concepts did require a little work to comprehend how to be successful in their application, and we learned that these other organizations were not willing to invest any effort in learning anything new.

As the ISA-88 concepts became institutionalized for batch processes, an interesting thing happened where there were both batch and continuous processes in manufacturing that had to work together. Having different approaches to delivering automation in a single environment created its own "pain," and the resolution that had the least pain associated with it was to select the ISA-88 concepts of modular equipment control and recipe procedural control and apply them to the total manufacturing process. The results were astounding, and those who were part of these efforts would attempt to carry the concepts back to their organizations, often with great success, for a while. This is where the organizational momentum and institutionalization of approaches that had been proven to be inferior surprised those of us in the batch segment of manufacturing. Over time spans that measured only a few years, the champions who were attempting to improve the continuous and discrete industries were moved out and the inferior institutionalized approach reasserted itself. To be fair, without the organizational support and acceptance of the concepts, adoption of anything new is doomed to failure. Business management seems to be clueless to the differences in approaches, so it is left up to the engineering disciplines to sort things out, which generally means "don't mess with success or take risks that might jeopardize a career."

This brings us to an interesting time in the early days of the new century. The packaging industry was struggling with issues that were requiring it to rethink how it delivered automation with packaging machinery. Working through the OMAC Packaging Workgroup (OPW), an effort to improve that industry was launched and championed by several large influential end user companies. This effort progressed with several deliverables; one of the most notable was the PackML guidelines Version 2 and Version 3, which provided guidelines for some very successful applications. Also during this time the ISA88 committee was evaluating the need to form a new working group to add more clarity on how to deliver modular equipment automation. During a meeting in Orlando in February 2004, Rob Aleksa of Procter & Gamble approached the WBF and suggested a joint group be formed between the batch and packaging communities to evaluate if it was possible to jointly deliver something that would benefit both communities.

Thus the Make2Pack initiative was born, which I have led. (I started this as a Procter & Gamble employee and have continued to lead the effort after retirement.) This effort took over 2 years to complete with the investment of over 5000 effort hours involving 180 companies and 250 individuals. The results were a resounding affirmation that the two groups should work together and jointly create a global standard that would support batch and packaging. The ISA88 committee, under the leadership of Dennis Brandl, chartered the "Batch Control Part 5: Implementation Models and Terminology for Modular Equipment Control" working group, which got to work in 2006. During this effort, the group of packaging and batch practitioners identified and started work on several necessary additions to the effort that were outside of the scope of the charter. After reviewing the expansion to the part 5 charter, the ISA88 committee directed the part 5 group to refocus their efforts and remain within the original charter. They also authorized the part 5 group to develop a technical report that would address the needs of the packaging members of the committee.

The ISA technical report "TR88.00.05 Machine and Unit States" (an implementation example of ANSI/ISA-88.00.01) was started in 2006 and delivered to ISA in July 2008 for publishing. At that time ISA determined that the numbering was in error and changed the report to TR88.00.02. The report focused on how to use ISA-88 concepts to modularize a packaging machine and how to provide a new concept "Mode of Control" that provides multiple methods for managing a machine. The previous PackML guidelines only dealt with one Mode of Control, which was referred to as "Automatic." Using this term created a significant amount of misunderstanding in the early Make2Pack effort, as the term was used very differently from the way it is used by ISA-88. The technical report attempts to provide clarity and reduce confusion by adopting the term "Production" as the mode of control and not "Automatic." How helpful this will be remains to be determined.

Another milestone event during this time was that ISA-88 part 1 had been successfully applied for 10 years. As is customary with ISA/ANSI standards, 10 years after acceptance the standards are evaluated for continued relevance and, if relevant, for any updates that might be required. This evaluation led to an update effort that was led by Paul Nowicki, which was proposed to take only a few months. In the end, after several years, the part 1 update looks to be ready for acceptance. With the parallel efforts of part 5 and part 1 being so interactive and in the end dependant on one another, the decision was made by the new ISA-88 chairman, Randy Dwiggins, to suspend the part 5 work and focus on making sure part 1 addressed any and all contentious issues part 5 had identified, which the part 1 update group has done. With the part 1 effort reaching completion, the part 5 effort is again progressing and will deliver proven automation approaches based on ISA-88 concepts that can benefit all manufacturing industries.

The adventure continues!

David A. Chappell
August 2010

ISA-88 Isn't Just for Batches Anymore

Presented at the WBF
North American Conference,
March 5–8, 2006, by

Robert Zaun
rmzaun@ra.rockwell.com
Rockwell Automation
6950 Washington Avenue South
Eden Prairie, MN 55344, USA

Adam Maki
abmaki@ra.rockwell.com
Rockwell Automation
6950 Washington Avenue South
Eden Prairie, MN 55344, USA

Abstract

Applying the ISA-88 standard to continuous and discrete applications can save time and money. The ISA-88 modular concepts that lead to a structured design approach have gained a lot of momentum by providing cost savings in the batch process and packaging industries. Are batch processing and packaging the only industries that can benefit from standards and models? This chapter suggests that all control systems can benefit from the ISA-88 standard approach. System design and startup time reduction, easier long-term maintenance, and overall lower cost of ownership are benefits of a standard approach that apply to a wide variety of control applications. In conclusion, this chapter illustrates the potential time and cost savings attributable to applying the ISA-88 standards and models beyond batch processing applications.

Introduction

At the dawn of the ISA-88 standard, it was recognized that standard models and terminology were required before various batch control vendors and users could communicate. The lack of a standard resulted in many varied solutions to the batch control problem. These custom solutions were only understood by the individuals developing them. The ISA-88 standard has provided a vehicle for uniformly implementing batch solutions on different processes, at different sites, and even in different industries. By following the ISA-88 standard models, anyone familiar with the standard is able to look at any code and have a general idea of how it is structured and how to modify it if necessary.

The same issues exist in the continuous and discrete manufacturing industries. The Make2Pack effort recognizes this fact and is well on its way to finalizing conceptual models and methodologies that apply to the total manufacturing process. In addition, the ISA-88 committee will be issuing part 5 with the goal of applying ISA-88 to continuous and discrete processes as well. This chapter will illustrate how applying ISA-88 concepts can improve your discrete processes today.

How Can a Batch Standard Possibly Work in a Continuous or Discrete Application?

Whether one is looking at the physical model of batch, continuous, or discrete manufacturing, similar levels of equipment can be identified. At the lowest level there are simple devices such as valves, switches, and so on. These simple devices typically group to perform a higher level of functionality, and this higher functionality will ultimately produce a product. In the batch world, these are known as Control Modules (CMs), Equipment Modules (EMs), units, and process cells, respectively. Similar divisions can also be made in the continuous and discrete worlds.

The WBF Make2Pack effort has already begun to correlate terminology between batch, continuous, and discrete manufacturing. As seen in Figure 1.1, a slight change in the ISA-88 terminology makes it transferable to these other realms of manufacturing.

Why Use a Modular and Structured Programming Approach with Common Definitions?

Using common definitions and methodology across the batch, continuous, and discrete operations required to make a product eliminates confusion when translating

S88 part 1 term	Make2Pack term	Make2Pack Description
Process Cell	Production Line	A collection of one or more machines, linked together, to perform one or multiple tasks of the process for one or more products in a defined sequence. • Continuous Process (e.g. forming line in the food industry) • Converting Line (e.g. paper, fibers) • Discrete Manufacturing (e.g. assembly) • Packaging Line (from filling to secondary and tertiary packaging) *S-88: A process cell contains all of the units, equipment modules, and control modules required to make one or more batches/lots.*
Unit	Unit Machine	In packaging, the unit corresponds to the logical grouping of mechanical and electrical assemblies that historically have been called machines. The term unit may apply to single function machine (filler, capper) or a multifunctional machine (monoblock filler/capper or any other configuration that combines functions within a single machine frame and control system). A multifunctional machine/unit can perform some or all of the functions of a packaging line, corresponding to process cell, that perform some or all of the functions of primary, secondary and tertiary packaging. A multifunctional machine may be logically broken down into several units corresponding to the individual functions. *S-88: A unit is made up of equipment modules and control modules. The modules that make up the unit may be configured as part of the unit or may be acquired temporarily to carry out specific tasks.*
Control Module	Control Module	A Control Module is the lowest module in a physical model breakdown of a unit. The term Control Module relates to the combination of (a) physical equipment and the lowest level control component that controls this equipment to carry out a physical process action. There may be control modules without directly associated physical equipment. These control modules coordinate/supervise/sequence other control modules. NOTE: The use of the term control module to describe the supervisory/sequencing/coordinating functions is proving confusing and difficult to convey the concepts. Needs further consideration. Make2Pack examples of Control Modules: • Servo • Conveyor • Pneumatic Cylinder with feed-back *S-88: A control module is typically a collection of sensors, actuators, other control modules, and associated processing equipment that, from the point of view of control, is operated as a single entity. A control module can also be made up of other control modules. For example, a header control module could be defined as a combination of several on/off automatic block valve control modules.*

Figure 1.1. ISA-88 and Make2Pack terminology.

meaning from one application to the next. A good example of the benefits of common definitions can be found in the following military example. If one branch of the Norwegian Military were to pass the command "secure the building" on to another branch, they would each interpret that command differently. The army would surround it, provide cover fire by snipers and heavy machine guns, and then clear the building room by room. The air force would make sure the windows were closed, the lights turned off, and the doors properly locked at the end of the day. The navy would sign a 10-year lease contract, with options to buy it after 5 years. Each of these actions are appropriate and meet the needs of the individual branches of the military but could bring disastrous results if the command was given from one branch to another.

The same is true for the steps in a manufacturing process. To improve manufacturing output as product is passed from one production step to the next, information must flow faster and transition times must be reduced. As companies continue to connect the top floor to the shop floor, "secure the product" will need to mean the same thing in every step of the process. When the business management system sends a command to the manufacturing floor, all processes and machines must have the same understanding of that command to ensure that the command is executed properly and safely.

Demands on Manufacturing

What demands are placed on today's manufacturing facilities? The same that have always been—today's manufacturing facilities are asked to provide an ever-increasing return on investment for the products they manufacture. To achieve this end, the team is challenged to continually make product faster, better, and cheaper. How does following the ISA-88 standard models help reach these goals? Let's examine each individually:

1. Faster

- *Reduced design and startup time.* By reusing modules of code, design time and the time required to start up and troubleshoot a new machine are reduced.

- *Greater process uptime.* When consistent code is used throughout the manufacturing processes, operations and maintenance workers can more easily find and understand the root cause of a maintenance issue.

2. Better

- *Improved quality*. Using standard terminology to define the components and processes involved in manufacturing, Quality Assurance is able to define consistent metrics for continuously improving the quality of the product.

- *Improved validation*. Common code reduces the time required to validate changes, as a change can be validated once and used in many different places.

3. Cheaper

- *Less initial capital investment*. Common code modules allow manufacturers to reuse code, enabling shorter design cycles and startup time and ultimately less up-front investment.

- *Less training*. Since the same concepts and code are used, operators and maintenance personnel do not need to be retrained every time a new process or machine is installed in their plant.

- *Improved efficiency*. Consistently defined states throughout all modules of equipment allow for quick identification of which modules are not being fully utilized.

Applying ISA-88 principles can help achieve the continuous improvement goals that all manufacturing facilities must reach. The next section illustrates such improvements when applying ISA-88 standards to non-batch operations.

A Real-World Example of a Discrete System Using ISA-88 Standards

The following is an example of implementing a discrete machine using the ISA-88 standards and concepts. The application is from the packaging industry and is known as a Vertical Form, Fill, and Seal (VFFS) machine, as illustrated in Figure 1.2. The machine consists of four main operations: weigh, bag form, feed roll, and seal and cut. In other words, the machine weighs out some product, forms a bag or package around the product, feeds it through, and cuts and seals the package.

The Original Equipment Manufacturer (OEM) broke the machine down into EMs and CMs as suggested by the ISA-88 standard. As illustrated in Figure 1.3, each of the four main operations of the machine was configured as an EM. The elements and devices within each EM were defined as CMs. The entire machine

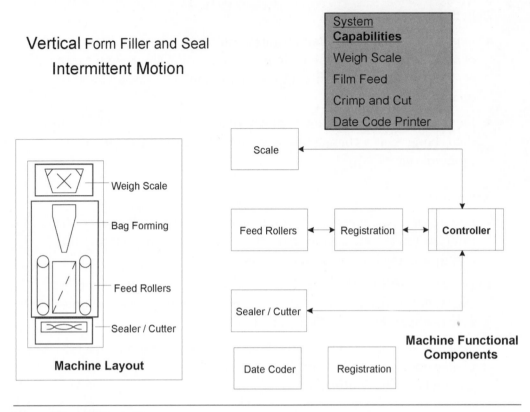

Figure 1.2. VFFS machine.

was defined as a unit. By defining the machine in this way, the OEM was able to create modular code that could be reused within this machine and also on other applications.

The OEM was then able to structure the code into modules that could utilize a modified ISA-88 state model as the foundation. This made the code easier to write and understand because there was a model to follow. The unit (machine) program transitioned according to rules set forth by the state model. The state model could then be used to develop a functional specification, programming document, training document, user document, and Human Machine Interface (HMI).

Program Design Methodology

The program design methodology consisted of the following:

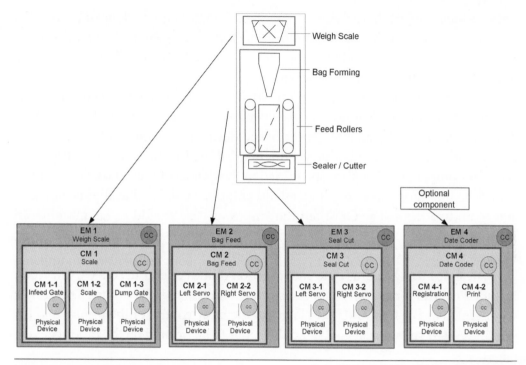

Figure 1.3. VFFS EMs.

- Define Physical Model
  - Break the unit (machine) into Functional Components and Sub-components (EMs and CMs)
- Define Equipment Procedural Model
  - Drive the complexity as low as possible into the CMs
  - Make interface to CMs as simple as possible
- Populate program

Program the CMs and EMs

With the Physical and Procedural Models defined, the OEM programmed nine CMs for this machine. Five of the modules were servo devices, so the OEM was able to duplicate the code for these, which simplified the program structure. By following this structure, the OEM and the end user only needed to understand one servo device module as opposed to five modules (each coded slightly differently because they did not share common states and functionality as defined by ISA-88). The OEM used a specific algorithm to start, stop, reset, gear, and cam the servos. With one CM being defined for servo devices, the OEM was able to consistently apply this algorithm to all of the servos.

In addition, the scale vendor provided some basic control code that was designed to ISA-88 principles (the scale vendor also supplies the batch industry). This code easily "plugged into" the code and design philosophy of the machine as an EM with three CMs. The scale vendor defined the functions of the scale system with the complexity of its function buried in the details of the "black box" control code. The interface was easily understood and clearly defined as inputs, outputs, status, and so forth. The OEM felt the scale was easy to add to his machine. In reality, a scale may be extremely complex, with multiple scales and an averaging algorithm. All this was transparent to the OEM.

The OEM designed the date coder section of the code from scratch. This code included control for one servo for registration correction and a printer. The printer was defined as the final CM, with logic for the basic control of the printer and for sending ASCII strings that contained the codes to be printed on the product.

Define Linkages between Modules

The OEM next considered the linkages between the base CMs and their respective EMs. In the servo devices' CM there is interlocking for electronic camming and other servo-only level functions that are in the CM code but not exposed to the end user. The cam timing and parameter values can be accessed, but access to the algorithms for starting and stopping is of no value to the end user. The end user of the machine may only care about the system starting and stopping smoothly and the cycle time. The OEM can do its magic by delivering a machine that performs optimally within the user guidelines. By creating cam profiles that maximize the acceleration and deceleration within the required product cycle time, value may be added to the machine by delivering smoother operation and lower shock on the machine and product.

The OEM considered how the CMs would be accessed and what data would be transferred in and out of the module. In this way the interface to the CM was as clean and intuitive as possible.

This effort made "plugging" the CM code into the EM simple because the interface points were clearly defined. By coding this module and testing it in one application, the OEM was confident that the code would work when plugged into other parts of the machine or into other machines.

Programming the Operation of the Machine

At this point the code was set up to represent the physical structure of the machine. Next, the OEM looked at the operations that the VFFS machine needed to perform. In this case the machine needed to do the following:

- Thread up the machine with packaging materials
- Start up production of the machine
- Empty out the machine
- Manually jog one section

When designing the control code for a machine using ISA-88 concepts and methodologies, at the highest level of machine control one considers the physical and procedural models and how they interact together in a machine to provide useful functions. From an ISA-88 perspective these functions correspond to Unit Procedures or Operations for the machine. The OEM set up the functions of a machine as phases. From this level, the OEM considered which phases of the machine to invoke. An individual phase such as threadup or many phases may run simultaneously (or as a recipe) while producing a product.

Benefits Derived from Using the ISA-88 Standard

By following the ISA-88 standard, the OEM would be able to reuse this code for next-generation machines or other packaging applications. For example, the OEM was able to use the weigh scale EM without modification for a different machine that also included a weigh scale. Likewise, the OEM would be able to reuse the servo device CM and printer CM when appropriate. Codes that are structured to fit the ISA-88 models are powerful because they can be separated at different levels (i.e., unit, EM, and CM) and plugged into other applications where appropriate because of their common definitions and structures. This avoids large blocks of monolithic code that are difficult to understand and even harder to debug and maintain.

The OEM has also found that their field support personnel are now able to quickly understand the machine code and have expressed appreciation, as the modules have found their ways onto other machines. The field support personnel no longer have to understand how control engineer Frank programs versus control engineer Ted. Now Frank and Ted use the same structure by following the ISA-88 principles. If Frank and Ted were authors, they would both write their books with the same structural components: title, table of contents, chapters, paragraphs, and sentences, and possibly even an index and a glossary.

From a maintenance perspective there is a benefit for both the OEM and the end user. By following the standard, the customer is able to easily understand the structure of the code, eliminating the time required to troubleshoot issues when they arise. In addition, if the customer performs batching operations they may have other ISA-88 based installations within their facility. Since the code is easy to understand, the customer is less likely to make a mistake when modifying the code, which minimizes the support costs for the OEM. If there is an issue it should be easy to locate, as the module of code associated with the device in question is defined as a specific CM.

In an ideal world the end user would never need to modify code once a machine is delivered; however in the real world one must be prepared for such an event. Using models and terminology as defined by ISA-88 was key in helping this OEM successfully manage changes.

As this OEM continues to roll the ISA-88 standard into other applications, it believes it will reduce design time, startup time, installation time, support trips, and end user training time. These savings will significantly reduce its overall cost of designing a machine while improving the quality of the end product. This standard allows the OEM and its end users to meet the goal of making the product faster, better, and cheaper.

Conclusion

The previously stated examples illustrate real improvements and cost savings to a discrete manufacturing process by implementing ISA-88 principles. In addition to the immediate efficiency gains on the manufacturing floor, a consistent and commonly defined programming structure enables a connection to business-level systems by enforcing the use of common definitions and methodologies.

One can see that the ISA-88 standard isn't just for batches anymore. ISA-88 concepts can have the same value for continuous and discrete processes as well. Future efforts from the Make2Pack and ISA88 committees will continue to evolve and improve the ability to implement structured programming concepts and methodologies. These efforts will provide a standard and consistent programming structure across automation at the process, site, and enterprise levels.

Applying ISA-88 to Non-stop Production

Presented at the WBF
North American Conference,
May 15–18, 2005, by

Dennis Brandl
President
dnbrandl@brlconsulting.com
BR&L Consulting, Inc.
208 Townsend Court
Suite 200
Cary, NC 27511, USA

Introduction

While the ISA-88 standard has been successfully applied to many non-batch problems, there is no consistently defined method for applying it to non-stop production. Non-stop production is defined as a continuous product flow through a facility, with no breaks in product flow even when products change. Non-stop production is also typified by small units that perform actions on part of a batch, such as filling a few bottles of a batch at a time, applying a coating to a product as it moves through a specialized coating machine, or packaging a product into a box one at a time for each product in the batch. Non-stop production may be continuous, such as in the production of tobacco, or discrete, such as in the creation and filling of bottles or vials. An extension to the ISA-88 model has been defined, called Non-stop 88 (NS88), which allows standard batch execution engines to be applied to these processes, resulting in simple recipes and bringing the power of recipes to continuous and discrete non-stop production. This chapter defines NS88, demonstrates how it provides a reliable Programmable Logic Controller (PLC) and Distributed Control System (DCS) control structure for these problems, and discusses how it has brought measurable benefits to real non-stop control problems.

Non-stop Production

Non-stop production is used by both continuous and discrete manufacturing where there are no breaks allowed in the product flow. Non-stop production may be required because the physical process doesn't allow stops or breaks in products, such as in fiber-optic cable production. Alternately, non-stop production may be required when effective use of the manufacturing equipment means that line breaks between production runs and even different products are expensive and must be avoided, such as in filling and packaging operations. Non-stop production occurs in both continuous industries and discrete industries. It is a unique feature of these industries and is a different method of production than the one described in the ISA-88 Batch Process Control standard. However the ISA-88 models can be applied in non-stop production with only minor changes to the ISA-88 concepts and some specific requirements for equipment control. The NS88 model has been verified in a large, continuous product flow system with stringent non-stop production requirements.

The ISA-88 model has become the preferred model for batch production systems. In these systems a production batch will move through a set of equipment as it is produced, but it follows the rule that there is only one batch in a unit at a time. At the time of ISA-88's development there was no attempt to apply the standard to continuous or discrete manufacturing, but the standard does mention that the models and concepts may be applicable.

There are many continuous and discrete manufacturing problems where there is no commonly accepted control system model. Each solution seems unique and there appears to be no underlying pattern that can be applied from job to job. Usually these problems occur when there are multiple paths through the production equipment, either through conveyor arrays or valve arrays, and where the process can be modified by sending execution parameters to equipment.

Discrete production examples include the movement of products such as electronic boards, cooked and precooked food products, consumer products, or bottles in a continuous manner through a set of processing steps. Continuous production occurs when the product moves in a flow and there are no discrete countable elements. In continuous production the product moves in a solid flow (e.g., liquid mixing or cable assembly) or as a set of undifferentiated small elements (e.g., fruit sorting and tobacco production). Continuous production examples include beverage mixing, chemical production, and various food processing. Examples of execution parameters include systems with dryers, where the drying temperature and dwell time may vary by product, or systems with fillers, where containers are filled with different products (e.g., apples in apple pie, soda in bottles, drugs into bottles, and bottles into boxes) based on equipment parameters.

In non-stop production there can be multiple paths, multiple products on the line at the same time, and multiple equipment configurations. In these and other cases it is often too expensive to have breaks in the production line just because the control system is not smart enough to manage this complexity. The control system must be smart enough to switch products in the middle of the stream, keep track of where each product's production run starts, and track the information about each production run.

Engineers and companies developing control systems for these applications often come up with their own standards. Many solutions are based on years of trial and error with systems that work but may have little overall structure. Unfortunately, systems applied in one application are often hard, if not impossible, to apply in even slightly different applications.

Non-stop Production Lines

The aforementioned dilemmas or problems are similar to those faced by batch production before the introduction of the ISA-88 standard. Fortunately, the lessons learned from ISA-88 can be effectively and efficiently applied to the non-stop production applications. Slight extensions to the ISA-88 rules allow the models to be applied to non-stop production. In addition, these same extensions mean that existing ISA-88 products can be used in "stop allowed" continuous and discrete production.

Let us look at a diagram derived from a real example, as illustrated in Figure 2.1. In this application each box represents a specific process applied to the product. These can be heating, cooling, inserting ingredients, wetting, soaking, drying,

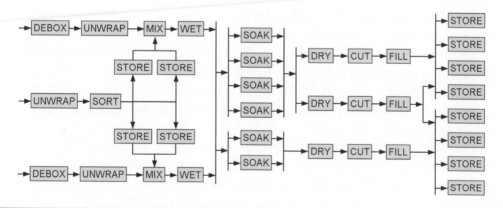

Figure 2.1. Non-stop continuous production.

filling, capping, labeling, storage, testing, and washing. The arrows between the boxes represent conveyor sections that move the product between units. The places where there are multiple different inputs or outputs on the conveyor sections indicate conveyor switches or conveyor arrays. Some of the processes, such as soaking, may operate on a whole production run at once; others operate on a stream of product.

If you are familiar with the ISA-88 models, the diagram should look very familiar; it is a unit layout diagram. In a typical ISA-88 system, each box represents a unit, and the arrows indicate material transfers between the units. This diagram is not a Piping and Instrumentation Diagram (P&ID); it is not a process flow diagram but is instead a diagram of the equipment hierarchy and equipment connections.

At this point the exact product is not important in determining the design pattern. Many continuous and discrete processes could be documented using similar diagrams. The goal of NS88 is to define a set of rules that allow a common model across a wide range of processes.

Can ISA-88 Be Applied?

The obvious question is, can the ISA-88 models be applied to this type of production? Another equally important question is, can it be implemented using commercially available batch control systems? Fortunately the answer to both questions is yes. Applying ISA-88 to non-stop production requires changing some of the ISA-88 rules but not breaking the basic model. For purposes of this chapter the extended rule set is identified as NS88 (Fig. 2.2). This is a model that can be applied where product moves in discrete elements or as a continuous flow—but only where all the product may never be in a single unit at the same time.

The goal of NS88 is to maintain the powerful concept of separation of product definition (in recipes) from intrinsic equipment capabilities. This allows for changes in products without requiring changes to equipment control programs. NS88 also addresses the ability to have multiple products in the same process cell at the same time. The key change from ISA-88 to NS88 involves the ISA-88 rule that a unit only contains one batch at a time; in NS88 a unit is only assigned to one batch at a time.

There are also specific rules that equipment phases must follow. In Figure 2.1, each box represents a unit, but they are small units compared to the units in a typical batch system. These units, however, typically perform only one or a small number of process actions. For example, a unit may wet a product, apply a coating

A unit is only assigned to one batch at a time.

- A unit completes its main processing phase when the last element of the batch reaches the front of the unit.

- A unit keeps control of the old batch as it moves through the unit using the old batch's processing requirements.

- A unit keeps track of the last element of the old batch as it moves through the unit.

- A unit sends an end-of-batch to the downstream unit when the first element of the new batch reaches the end of the unit.

Figure 2.2. NS88 rules.

to a product, insert components into a product, or mix multiple product streams. Typically there can be one main operational phase for each unit, such as Wet, Coat, Insert (component), or Mix.

The main operational phase should include the following rules:

- The main operation phases must complete when the last element of the batch enters the unit. The next product to follow is designated as the new batch.

- Each unit or Equipment Module (EM) must continue to process the old batch still in the unit and keep track of where the batch boundary is within the unit.

- Optionally, each unit or EM may send an end-of-batch signal to its downstream unit when the last element of the old batch leaves the unit.

Applying NS88

The following example illustrates production using the NS88 rules defined previously. Figure 2.3 illustrates a single stream production line with two units. FILL #1 fills the product with material. COAT #1 coats the product with a material.

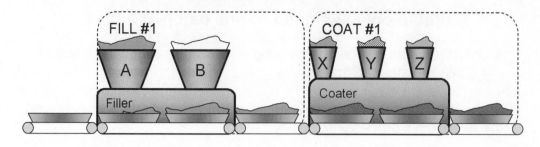

Figure 2.3. Sample non-stop production line.

Each unit has a main processing phase. FILL #1 has an equipment phase, Fill, and COAT #1 has an equipment phase, Coat. These phases have parameters that specify which material to use in their process.

The units include the conveyors section downstream from the main processing unit. The conveyors run from FILL #1 to COAT #1, with the assumption that there is something downstream from COAT #1 to catch and store the product. More complex conveyor systems are discussed later on in this chapter.

Figure 2.4 illustrates a sample recipe for this production line. This is a collapsed recipe with phases within unit procedures and no operations. Notice that there is no unit-to-unit transfer in this recipe. Using the NS88 model, unit-to-unit transfers are not required between continuous units. This means that NS88 recipes are simpler than ISA-88 recipes because they are constrained by the movement allowed by the physical equipment. This is a very simple example, but it illustrates how the NS88 rules for unit assignment and phase execution allow for a non-stop switch of product in continuous production.

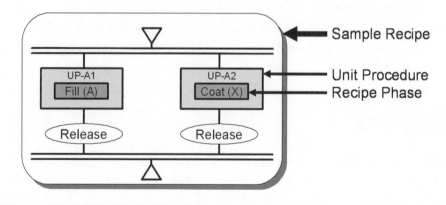

Figure 2.4. Sample recipe for a non-stop process cell.

Batch B2 Starting Fill

A fully detailed example of NS88 is too complex for this chapter, but a look at the system in the middle of a unit switchover illustrates the application of the rules. The situation is as follows:

- The first unit (U1) of Batch B2 has entered FILL #1
- U1 is assigned to B2, but the unit code is still keeping track of B1's location so it knows when B1 leaves the unit
- FILL #1 continues to fill the old batch with product A but is also filling the new batch with product B
- COAT #1 is still assigned to Batch B1
- The state of the system should be visible outside the batch system in a Human Machine Interface (HMI) or SCADA system, where FILL #1 would be coded to indicate it is assigned to Batch B2 but still has not yet cleared Batch B1, as shown in Figure 2.5

Generating the Right Batch Record

In the normal ISA-88 models, the batch record contains all the actions performed on a batch. There is only one batch at a time in a unit, the phases in a unit procedure are specifically assigned to a single batch, and the batch log can be organized by phases. In NS88 there are times where the unit contains more than one batch, and the equipment phase for a batch completes before the batch completely leaves the unit.

In cases where complete information about a batch must be defined in a single batch record, the following method can be used:

1. Reporting units should be created that match the processing units
2. The reporting units should contain phases that adhere to the following rules
   a. The report unit phases must complete when the last element of the batch leaves the unit
   b. The report unit phases report on all elements of the batch processed through the unit
   c. The report unit phases must coordinate with the processing unit's operational phase to track information about the new batch through the unit

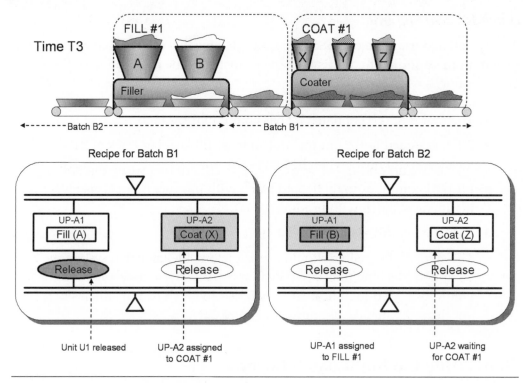

Figure 2.5. Time T3 product switchover in FILL #1.

These rules are summarized in Figure 2.6.

Figure 2.7 illustrates the reporting method. In the previously defined example, two new units would be created—FILL #R1 and COAT #R1. These units would have report phases that report on the batch still in the unit. From the batch execution system, they are units assigned and released in the same manner as the processing units. The recipe contains extra unit procedures for the reporting units.

Figure 2.7 illustrates the situation at time T3 in the example. In this case, the Report phase in reporting FILL #R1 has not yet been completed because the last of the batch has not yet left FILL #1. The batch log for Batch B1 would contain the reported information on all elements of Batch B1 in FILL #1.

When the last element of Batch B1 leaves FILL #1, then the FILL #R1 report phase can report the final information and then complete. FILL #R1 is then released and can be assigned to Batch B2.

This method doubles the number of units within a process cell and may double the number of phases, since the process phase (i.e., Fill) cannot return a report. In an ISA-88 system, the Fill phase may have two parameters that specify the material to fill and the amount to fill. The phase would also have a report

A unit is only assigned to one batch at a time.

- A unit completes it's main processing phase when the <u>last element of the batch reaches the front of the unit</u>.
- A unit keeps controls of the old batch as it moves through the unit using to the old batch's processing requirements.
- A unit keeps track of the last element of the old batch as it moves through the unit.
- A unit sends an end-of-batch to the downstream unit when the first element of the new batch reaches the end of the unit.
- Create reporting units that match the processing units.
    - The reporting units phases complete when the <u>last element of the batch leaves the unit</u>.
    - The report phase reports on all elements of the batch processed through the unit.
    - The report unit's report phase must coordinate with the processing unit's operational phase to track information about the new batch through the unit.

Figure 2.6. NS88 rules with reporting.

parameter that is returned by the phase with the amount of material actually filled. In the NS88 model, the Fill phase would not have a report parameter, and a Report phase would return the actual number of product elements filled.

Other Special Rules

The same model can be applied to starting up and shutting down the production line or process cell. Recipes can be used to sequence equipment startup, which is usually a requirement due to electrical load considerations.

Switching conveyor systems can also be handled using the same set of rules. For example, a switching conveyor system would be defined as a unit. It would have a main operational phase that directs its material flow down the proper path but would also control the previous batch down its path until it leaves the conveyor system.

Generally a production line is designed with some units that can buffer to handle temporary stops of equipment, such as that shown in Figure 2.8 in a continuous filling process. In this case the hold rule is applied when a hold signal is

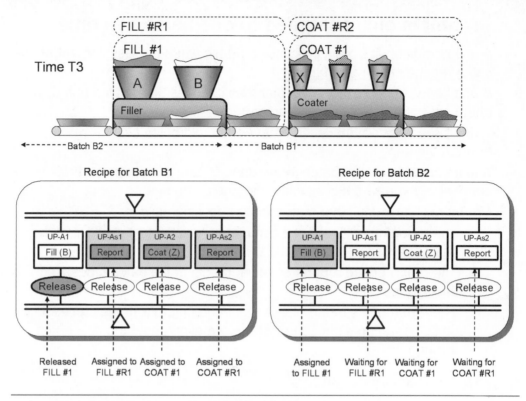

Figure 2.7.   Using reporting units to report on a batch.

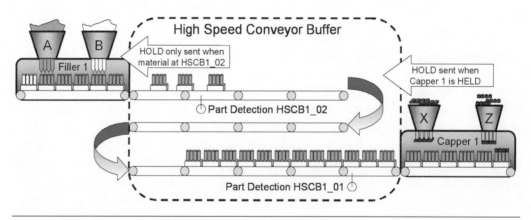

Figure 2.8. Buffering in a discrete process.

sent from a downstream to an upstream unit. This can be a peer-to-peer communication, or it can be handled through a batch execution system.

The high-speed conveyor system could either be a unit (when the hold signal goes through a batch system) or an EM (when the hold signal speed requires direct communication to the upstream and downstream phases).

NS88 should use a modified state model such as the one illustrated in Figure 2.9. Each unit and EM may have a different state model, but Figure 2.9 is a typical example of an NS88 unit, work cell, or shared EM. This model is derived from the Organization for Machine Automation and Control (OMAC) state model and includes extra states for product switchover used by NS88.

Equipment states are generally not the same for all pieces of equipment, in that the exact functions to be performed in each state will change based on the manufacturing process and equipment capabilities. Table 2.1 lists some example equipment state definitions.

Summary

The ISA-88 model can be effectively applied to non-batch systems and in particular to non-stop continuous and discrete production systems with only slight

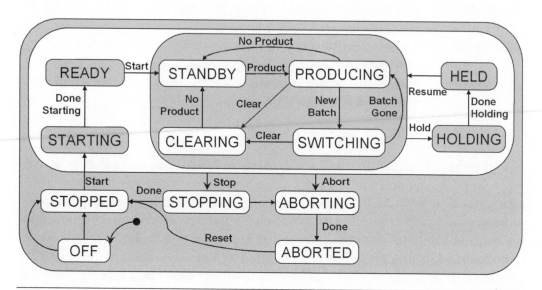

Figure 2.9. Example of a typical equipment state model.

Table 2.1. Equipment state definitions	
Equipment state	*Description*
OFF	The equipment is powered up, but is not yet ready to be operated. (It may be performing initial self tests and power up logic.)
STOPPED	The equipment is ready to be started. It is not performing its basic processing function and is not processing material.
STARTING	The equipment has been commanded to start and is in the process of starting, but the equipment has not yet been commanded to process material.
READY	The equipment has finished starting and is ready for operation.
STANDBY	The equipment is waiting for material to process.
PRODUCING	The equipment is performing its basic processing function on one batch.
SWITCHING	The equipment is finishing processing a batch and is starting to process a new batch.
CLEARING	The equipment is finishing processing a batch, and no new batch is present.
HOLDING	The equipment has been commanded to hold and stop processing the batch and is performing the actions to go to the HELD state.
HELD	The equipment is not processing the batch and has sent a hold signal to its upstream equipment.
STOPPING	The equipment has been commanded to stop processing material and is performing actions to clear the material.
ABORTING	All motion is stopped now, before something breaks or product is damaged.
ABORTED	The equipment is not ready to be started. It is not performing its basic processing function. It needs repairs.

modifications. The modified rules identified as NS88 can be applied to a wide range of problems and allow existing batch products to be effectively used for non-batch problems. NS88 changes the one-batch-per-unit rule of ISA-88 to a one-batch-assigned-to-a-unit rule. NS88 then provides an underlying pattern for the organization and state models for EM (programmed code) and for unit-to-unit communications. NS88 has been verified in a large continuous product flow system with stringent non-stop production requirements. In addition, the NS88 architecture was retrofitted onto an existing physical process with minimal disruption and changes to the underlying physical equipment.

Non-stop 88 Delivers Improved Operational Excellence

Presented at the WBF
Make2Profit Conference,
May 24–26, 2010, by

Mike K. Williams
Asset Manager
williamsm@dow.com
The Dow Chemical Company
1319 Building
Midland, MI 48667, USA

Andre Schepens
Process Automation Discipline Leader
The Dow Chemical Company
Terneuzen, Netherlands

Dave Huffman
Marketing Manager
ABB
Wickliffe, OH, USA

Abstract

Does the design of your process automation solution contribute to operational excellence? Do you wish to prevent business interruption due to detectable and preventable events? Does your business expect to deliver higher asset capability with increased first-pass prime production? Do you wish to improve regulatory compliance through mitigation of abnormal events? If so, state-based control strategies delivered in a modular manner could provide a low-cost, reliable solution to your business challenges.

Introduction

ISA-88 is a well-known standard within the batch automation community; however, the principles of batch operations can also be applied within the non-stop continuous process industries. Even continuous processes exhibit unsteady state behavior during conditions of startup, shutdown, and grade change. Furthermore, continuous processes also experience unsteady state behavior during unplanned processing events such as material plugging, equipment failure, or environmental change. It is during these events that the value proposition of ISA-88 offers a new opportunity for value creation.

Today in the capital-intensive process industries, return on net assets is of vital importance. This return can be expressed in terms of operational excellence metrics, including (but not limited to) asset utilization; mechanical reliability; environmental, health, and safety compliance; and first-pass quality, to name a few. The business desire is to sustain high operational levels or asset utilization, in which one accounts for every minute of every day's worth of production. To this end, abnormal conditions resulting in process downtime must be avoided if possible and mitigated when identified. The use of process automation solutions as mitigation tools can be extremely valuable and effective.

Tolerance of a process fault is not a new concept, especially as it relates to process control hardware. Redundant controllers, Input/Output (I/O), networks, and servers have been purchasable options for reliability reasons for many years. The concept of application fault tolerance to mitigate process upsets is a less-commonly deployed approach. The principle—be it applied in either a batch or continuous context—is for less-than-critical process disturbances, but the solution should detect the abnormal situation, alert the operator, select an alternative execution path, and automatically correct the deficiency based on preapproved and tested automated responses.

In many non-safety-related cases, it is satisfactory to run the process in a partially impaired state while the defect is being corrected by the operator or maintenance organization. For example, continuous processes install redundant physical equipment (such as dual pumps) to sustain operations in the event of the failure of one unit. In a duty standby arrangement, if a pump seal fails on the primary, this event can be detected with sufficient instrumentation or soft sensors, and a logical alternative can be taken automatically or manually with computer assistance, depending on the level of field instrumentation available. By computer-assisted expediting of the failed equipment scenario, the uptime of the process is improved thus resulting in reduced loss in production.

Given the sophistication and capability of modern process control computers, it is unclear why the technique of automating abnormal situational control has not been broadly applied. The answer to this question may be in the difficulty and cost involved with configuring such applications. Fault-tolerant applications with automatic disturbance handling require the construction of rather complex interlocking logic based on the detected process condition. This requires expert knowledge in both the process and advanced programming techniques. Consequently, such applications are expensive to construct and even more costly to maintain, debug, and enhance. A breakthrough is seemingly required to lower these barriers to implementation for abnormal situational management.

A solution is available in the form of state-based control for both batch and continuous processes using ISA-88 principles as the enabling methodology. To successfully construct a state-based application, the process state must be known to all control elements within the control strategy. In ISA-88, the concept of modes of operation has been employed to communicate process states within a control strategy. Examples of process states for a continuous process such as a distillation column could be maintenance wait, fill reboiler, establish total reflux, establish product takeoff, shutdown, and emergency shutdown. Given these modes of operation, one can program condition logic that provides alternative execution paths when a process changes from one process state to another. The condition logic can then be used to interlock or enable specific devices within the unit or equipment process state. The logic can be used to enable highly managed or smart alarms. It can also be used to automatically trigger procedural changes from one mode of operation to another.

A continuous distillation column is a generic example of a typical continuous unit operation that could benefit from a Non-stop 88 control strategy. In this case a common abnormal situational scenario could be a bottoms pump failure. Assume the column has been operational for several hours when a low pump seal fluid pressure occurs, indicating pending failure and a potential environmental release. Process computer detection of this failure notifies the operator of its preprogrammed intention to switch to a degraded rate of operation until the abnormal condition is resolved. The business does not wish to completely shut down the process, as they will loose 12 hours of productivity. Upon operator confirmation to the computer, the mode of operation, which is currently in overhead takeoff, moves to total reflux, thus keeping the process in a partially impaired yet ready-to-run process state. Once the column is repaired or the in-line spare pump is commissioned, the column, which is still at equilibrium, automatically returns rapidly to the fully on-line production state with minimal downtime incurred. Depending on the level of process equipment redundancy and associated instrumentation, this typical scenario can be mitigated in 15 minutes—a huge savings in lost asset capability.

In the previous example, communicating the process state condition between processing units is very important to the coordination of procedural activity, especially when integrating batch and continuous units within a process cell. In the real world, rarely does a process unit operation stand alone. Typically in a plant there are upstream, downstream, and utilities units whose actions are dependent on each other. In the case of fully integrated processes, communicating the state condition enables coordination control with smoother, sustained operations as a whole.

ISA-88 guidelines set the direction for developing process state enabled applications, even in the case of continuous processes. Non-stop performance can be achieved by applying modes of operation to logically linked Equipment Modules (EMs) and holistic units. Sequential Function Charts can drive the modes of operation with their process states, communicated within the complete processing unit using state agents. State agents can be employed to enable intelligent alarming and other forms of advanced procedural control (see Chapter 6).

Thirty Years of State-based Control

State-based control strategies have been deployed within the Dow Chemical Company for 30 years, using both proprietary and commercial process automation platforms. Recently, the aforementioned technique was applied during the retrofit of an existing specialty polymer plant. The chemical process in this plant was a combination of six batch and continuous unit operations close-coupled in series. The legacy system being replaced was a 25-year-old DCS configured with traditional loop control. At the heart of the process was a continuous distillation unit, which was the process bottleneck and historically the source of repeated product quality failures. Historically this unit experienced a high degree of unpredictable behavior due to variability in human response to unsteady state operating conditions. To remedy the problem, a state-based approach was taken in the new application configuration, where modes of operation were programmed for the process states, which were defined as startup, ramp up, running, degradation/ hold, and ramp down.

After interviewing the operations personnel and engineers using function analysis systems techniques based on past experience and Layer of Protection Analysis (LOPA), abnormal scenarios were identified with mitigating actions defined. These conditions were then programmed as operating actions executed during the degradation mode. Similar situational analysis was performed on other process operating modes to smooth transitions during commissioning and product grade changes. The programming effort was accomplished using a Sequential Flow Chart (SFC) language with visualization of process interlocks and transitions, provided automatically via preconfigured faceplates to facilitate debugging by the operations staff in real time.

Summary

The end result of this automation effort was a 15% improvement in plant asset capability and a two sigma improvement in product consistency. Payback on the automation effort was achieved in less than 1 year, with no change or amendment to the actual physical process. It has also been observed that there are fewer operator interventions, fewer alarms, and a lowered number of events reportable to the Department of Environment Health and Safety. This is primarily due to the preemptive detection, smart alarming, and mitigating actions achieved through automated modes of operation. Previously such events might have resulted in loss of primary containment of process contents. Now events such as tank overflow due to variability in operating procedure, planned and unplanned, have been reduced dramatically or eliminated completely.

The primary driving force for implementing state-based control applications within the Dow Chemical Company has been to enable business excellence. Concepts of Non-stop 88 have been successfully installed in both Greenfield and Brownfield cases, resulting in a significant return on investment based on operational excellence metrics. On average, state-based applications consistently generate a 200% or greater return on automation solution investments, based on the original Distributed Control System (DCS) replacement asset value. Moreover they are a critical part of the overall solution to achieve environmental, health, and safety compliance while balancing higher levels of asset productivity.

Further Reading

Huffman, Dave. Consider state based control. http://www.chemicalprocessing .com/articles/2010/051.html.

Seilonen, I., Appelqvist, P., Halme, A., and K. Koskinen. 2002. Agent-based approach to fault tolerance in process automation systems. Presented at the Third International Symposium on Robotics and Automation (ISRA 2002), Toluca, Mexico.

Taking ISA-88 Concepts to the Next Level

Presented at the WBF
North American Conference,
May 15–18, 2005, by

Jarrad Reif
Technology and Systems Manager
Jarrad.Reif@fostersgroup.com
Carlton and United Beverages
4–6 Southampton Crescent
Abbotsford, Victoria 3127, Australia

Gavan Hood
Senior Systems Architect
gwhood@ra.rockwell.com
Rockwell Software
156 Eggersdorf Road
Ormeau, Queensland 4208, Australia

Ralph Kappelhoff
General Manager
rlkappelhoff@ra.rockwell.com
Rockwell Software
1201 South 2nd Street
Milwaukee, WI 53204, USA

Abstract

ISA-88 is recognized as a successful model for modular control and production execution management in the batch industry. The value of the ISA-88 concepts spans beyond the process cell to all industry segments. This chapter presents an overview of opportunities for reuse and reapplication of batch processing concepts across downstream and upstream activities to the process cell. In addition, the authors present a vision of where the industry is headed in their representation of equipment control and recipe management in batch control systems

Background

ISA-88 concepts have proven to be successful in batch processing. Over many years the standard has withstood the tests of adoption, support, and time with a substantial percentage of implementations in the batch processing industry. Processes within the batch industry have evolved to represent implementations consistent with the standards presented in ISA-88 parts 1, 2, and 3. There are a number of other standards and guidelines to choose from in the industry. This chapter notes that the Organization for Machine Automation and Control (OMAC), ISA-88, and ISA-95 are converging to a consistent representation of manufacturing.

A number of groups have been formed to take the standard to new levels and perform valuable work on ISA-88 part 4's "production records" representation and ISA-95 part 3's "Activity Models of Manufacturing Operations Management," among others. These efforts have not been conducted in isolation. Members of these organizations are actively coordinating activities to identify overlaps and promote consistency in the standards efforts. In some cases these efforts have been formally recognized, with groups such as the ISA88/95 joint working group being formed to identify overlaps and improve consistency between ISA88 and ISA95 initiatives. Another collaborative effort, the Make2Pack joint workgroup sponsored by WBF, OMAC, and ISA has been chartered to develop better integration between "making" and "packing."

These activities are representative of a trend in major standards groups toward consistency across the standards. This will open up new opportunities for vendors that adopt the use of standards in their organizations and will allow companies to take their organizations to new levels based on a consistent standards base. Enterprise Resource Planning (ERP) vendors are realizing the value of integrated industry standards for manufacturing. For example, SAP is integrating support for the Business to Manufacturing Markup Language (B2MML) WBF implementation of ISA-95 parts 1 and 2 into their systems.

Challenges

Batch manufacturers are under increasing pressure to enhance efficiency and agility while maintaining or increasing flexibility in manufacturing across global systems. Efficiency and agility are expected from both new and already existing products and processes. This is a difficult challenge.

The ability to refine production processes is an art that requires a combination of expertise in process engineering and use of production information. For every achievement made in collecting and processing information, a new requirement is presented. Batch manufacturers must continue to push the boundaries available in technology using best practices. Typical challenges facing manufacturing facilities include the following:

- Deploying systems across all plant areas with consistent operating philosophies that minimize operator training and simplify the ability to move operators between roles. Systems should lead operators through the process and present clear information for decision making.

- Deploying systems across all plant areas with standard coding formats to minimize training requirements for maintainers and reduce engineering and commissioning time.

- Developing systems that expose diagnostics information so that programming tools are not necessary for troubleshooting operational problems.

- Capturing and aggregating process information from the control system and rendering that information so that accurate, real-time decision making can occur based on quantitative data.

Taking It to the Next Level

Proactive manufacturers take challenges such as these and turn them into opportunities to differentiate their system from the pack. By taking advantage of emerging capabilities in the standards communities and the vendors supporting these standards, proactive manufacturers are armed with tools to make the transition to a holistic (raw material to end product, Enterprise Resource Planning [ERP] to control system) integrated manufacturing system based on industry standards.

In Some Cases It Is Better to Do Nothing or Less

With so many standards in existence, standards committees must resist the temptation to create new standards and extensions without prior investigation of existing representations. In many cases, refinement of existing implementations or convergence with an existing standard can help avoid the duplication of effort and enhance synergy between standards. This reduction in duplication helps manufacturers converge on common, well-understood representations of manufacturing.

Integrated ISA-88/95-based Representation: End-to-End Tracking and Tracing

An integrated ISA-88/95 representation allows consistent use of components across batch, packaging, and warehousing; an example of such a system is depicted in Figure 4.1. This common representation allows consistent standards-based integration from batch to higher-level and adjacent systems. Figure 4.1 indicates what an end-to-end system representation will look like and provides a holistic system-wide representation of manufacturing with no boundaries between production areas. This holistic representation enables the goal of end-to-end tracking and tracing from raw material to end product (as indicated by the bold line) to transition from fiction to reality.

For many years software products based on the ISA-88 batch control standard have been available to the batch industry. These products have been well accepted, but like any system there are limits in terms of communications via phase logic interface with controllers. The next generation of process controllers

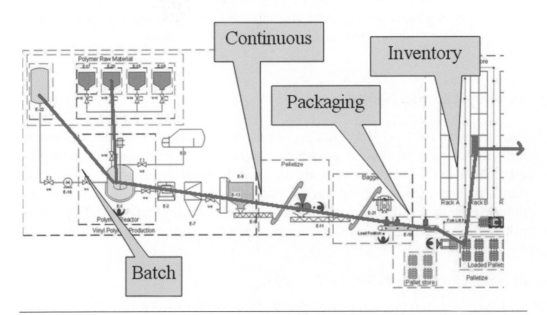

Figure 4.1. End-to-end system representation of manufacturing.

are representing equipment control elements (e.g., equipment phases and ISA-88 state machines) explicitly within the controller. The availability of these ISA-88 features in out-of-the-box controllers enables practitioners to more readily utilize ISA-88 modular control concepts across the manufacturing plant beyond the process cell.

Figure 4.2 illustrates the ability to code complex equipment sequencing actions directly into ISA-88 state machine interfaces shipped in the controller. This provides enhanced stability and the ability to easily execute time-sensitive actions. The ability to extend this capability to higher-level functions in controllers will provide more options for batch processing.

Figure 4.2 shows the representation of common terminology such as equipment phases and ISA-88 states within the phase (partially hidden by

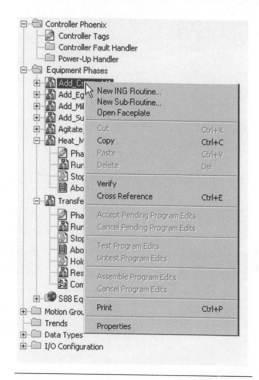

Figure 4.2. Example of standards-based controller logic.

the menu) for Run, Stop, Hold, Abort, and so forth. The ISA-88 based elements are integral with the controller development environment.

Application of Standards-based Control across End-to-End Production Processes: Batch, Packaging, and Warehousing

For many years, proponents of batch manufacturing have contended that ISA-88 modular control concepts are applicable outside the batch process cell. The representation of Control Modules (CMs) and Equipment Modules (EMs) across industry segments is something that should be pursued. OMAC has taken batch concepts into the packaging domain. The industry as a whole remains actively engaged in the formal adoption of modular control concepts described by ISA-95 and ISA-88 in other areas.

Figure 4.3 presents the equipment hierarchy spanning all industry segments. The presentation of EMs and CMs in areas such as inventory allows sharing of modular control concepts in a host of plant-wide systems. The generic model on

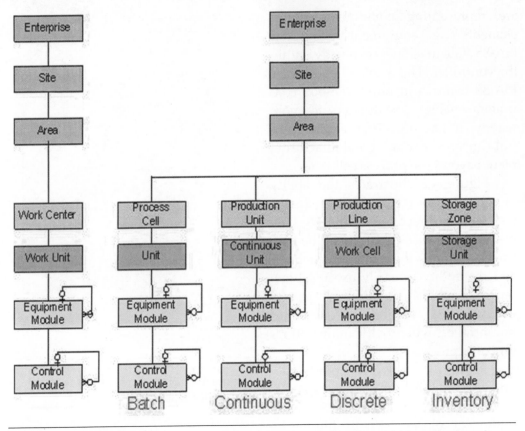

Figure 4.3. Generic and industry-specific representations of equipment hierarchy, based on ISA-95 part 3, draft 22.

the left has been included in ISA-95 part 3. This generic model helps us consider ISA-88 concepts in continuous, discrete, and inventory applications.

Example: Application of EMs to Conveyors

Conveyors are used in numerous industries to move items (e.g., cartons, bottles, and car bodies) from one place to another. In some cases conveyors are used to space items on a line so that another machine can perform the next process step on the product in a more efficient manner. Figure 4.4 shows a simple conveyor system, a "spacer system," that can be used to accomplish this.

Within the "spacer system" there are three conveyors:

- *In-feed conveyor*. Accumulates cartons for subsequent gapping.

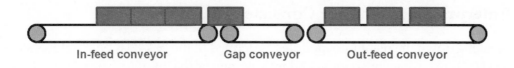

Figure 4.4. Conveyor configuration for gap generation.

- *Gap conveyor*. Ensures items have a minimum gap between them required by light sensors in the system. To achieve this gap, the conveyor must run at a faster speed than the in-feed conveyor, which will enable gaps to be generated.

- *Out-feed conveyor*. Moves gapped items to the next location.

The out-feed conveyor speed must be at the same speed or faster than the gap conveyor. The gap conveyor runs at a speed faster than the in-feed conveyor. Thus one can infer that the out-feed conveyor must be running faster than the in-feed conveyor.

We can manage this system using EMs and controls as represented in Figure 4.5. CMs manage the control of drive devices for startup and speed changes of the conveyors. The drive CMs are coordinated by EMs for each conveyor, which in turn are coordinated by the higher-level Spacer EM. The requirement of this conveyor system to coordinate the speeds of the three conveyors fits well with the application of EMs. EM-Spacer is responsible for coordination of the three conveyors and their associated EMs (e.g., EM-Infeed, EM-Gap, and EM-Outfeed). External systems can control the conveyor system via the exposed phases on EM-Spacer such as Start, Jog, Hold, and so on.

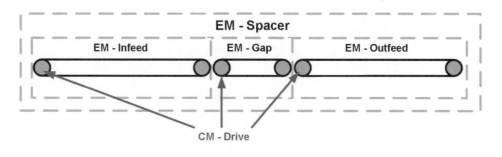

Figure 4.5. EM representation for coordination of conveyor system.

Enterprise Systems Are Consistent across Sites

Breaking down the plant into modules as defined in the ISA-88 models allows standard CMs to be developed independently as long as the interfaces are standardized. As a result we are able to develop different modules at different sites and then share these. The sharing of standard code modules allows numerous reviewers to incorporate improvements that are then incorporated into all plants—a practice that shares a similar philosophy with open source initiatives in the Linux world.

Having a consistent equipment hierarchy enables us to achieve better consistency and code reuse across manufacturers' facilities. Adoption of an industry standard model for modular control such as ISA-88 EMs and CMs enables vendors to better communicate with each other and allows reviewers to provide input about improvements for the system.

What Will This System Look Like?

The plant of the future may have the following:

- A global view of all processes within the plant via the corporate intranet

- Access to plant Key Performance Indicators (KPIs) in real time via the corporate intranet/portal

- Use of expert systems to monitor plant performance and highlight poor performance in real time

- Use of expert systems and plant models to optimize plant scheduling

- Use of real-time KPIs by operators to improve decision making

- Increased use of plant floor data to update corporate Manufacturing Execution Systems (MES) in real time, allowing MES to provide real-time insight of the supply chain

- Extension of tracking and tracing from transactional data at the Material Requirements Planning (MRP) level down to time series data at the plant floor

Opportunities

The consistent models presented by the convergence of standards and the availability of standards-based products in new forms open many doors:

- Reuse of advanced batch modular control skills across the enterprise
- Dynamic systems that can be downloaded and configured in standard modules
- New vendor–integrator–manufacturer relationships based on open systems
- Original Equipment Manufacturers (OEMs) presenting systems in standard, pretested, and understood configurations to manufacturers

More Auditable Systems

The production record initiatives identified within ISA-88 part 4 present an opportunity to provide more auditable representations of production information. To achieve this advanced level of tracking and tracing, production data should be stored in a manner that associates the data within a context that includes products being produced, locations within a production facility, and the time at which the data were generated. This transforms opaque production data into a valuable production information resource. The establishment of these associations often spans the entire system and multiple vendors.

The integration of the ISA-88 part 4 standard with ISA-95 will enable the integration of production records with higher-level systems in a fashion that is more consistent than what is currently possible. The support of vendors for ISA-88 part 4 will lead to the production of products that share a common base for representation of production records and auditing of systems.

More Consistent and Manageable Systems

Standardization of control system coding around known state machine representations produces the following effects:

- Allows the development of consistent implementations in all production areas including batch and packaging areas
- Leads to lower total cost of ownership, due to reduced training needs and common skills applied across the manufacturing plant
- Allows systems to be developed that will support all areas of the plant without customization to fit individual process areas, therefore minimizing implementation time and spreading the benefit of the development time

- Allows minimalist change and validation in modular systems capable of being managed at the component level, thus resulting in better application of change management

Summary

A new and exciting range of standards-based interoperability and convergence capabilities are nearing full development and implementation. The ability to move standardization of core batch processing capabilities directly into process controllers introduces a new technology to the batch industry.

These capabilities provide valuable tools for batch processors to leverage proven standards-based technologies into other areas of the manufacturing plant, reducing total cost of ownership of manufacturing through improved modularity and consistency across the manufacturing process.

The challenge to the batch processing industry is to review batch processing activities in light of these emerging and available capabilities and take their processing system to new levels of efficiency, manageability, and flexibility.

Experiences from Implementing a Single Control System Solution Using ISA-88

Presented at the WBF
European Conference,
October 11–13, 2004, by

Flemming Larsen
Principal Scientist
Methods & Systems
FlL@nne.dk
NNE A/S
Gladsaxevej 372
2860 Søborg, Denmark

Abstract

The world of the pharmaceutical and biotech industries is becoming more and more complex. Business production processes must be transparent in order to react quickly and accurately when necessary. Therefore engineering and project execution methods for production facilities have become more and more critical and demanding in order to achieve and realize safe pharmaceutical and biotechnology production.

This chapter will summarize experiences from a highly automated process plant (from the perspective of process control) with well-structured collaborative engineering executions, procurement, construction management and validation, and management methods to meet the aforementioned challenges. What seems simple or straightforward in theory (standards and international guidelines) can become extremely difficult in reality, but for a vendor of multiple process equipment projects with a single control system, it's extremely important to achieve precise project interpretation of standards like ISA-88 supported by well-structured process and automation-oriented planning.

In particular, this chapter places special focus on global testing activities and strategies based on ISA-88 principles, organized with multiple international process equipment skid vendors coordinated via one control system vendor.

Introduction

The pharmaceutical plant mentioned in this chapter is built in a modular fashion, with multiple process equipment suppliers being responsible for delivering completely tested and documented process modules with processing equipment. These process modules contain the total system including control systems with application software. The integrated and collaborative process automation engineering, planning, and the modular approach all created a fast-track project moving according to the customer need.

But many manufacturers of process equipment do not feel very confident with designing and taking the responsibility of full automation into fully functioning process modules. That is why this chapter will highlight how the use of structural front-end engineering decisions and designs that follow international standards and guidelines (and in particular the ISA-88 principles) can help to achieve standard methodologies for consistent systems. This provides major benefits for all involved in the project and creates confidence in method and technology at the process-oriented equipment contractor's organization.

The customer's approach for shortening "time to market" of the product was another high-level challenge for the project with a time requirement of 18 months from project execution start to handover. This required a strategic combination of technology, methods, and customized organization structures (including collaboration between multi-discipline engineering team efforts at different levels). One key issue involving the time savings was the ability to validate the software before automation equipment was installed, as opposed to after its installation.

A structured collaborative organization performed requirements, analyzing methodology to identify user expectation of the plant operability. This was one successful key tool for engineering and design execution and project management.

This chapter first deals with the project scope and then with the engineering, design, and execution from front-end decisions to handover. It also details the set up and use of ISA-88 Models and Structures, from the initial design phase through construction to validation, as well as the executed project management setup for achieving the target. Finally, the lessons learned will be highlighted.

This chapter excludes the basis of project decisions and instead will focus on the concept of the control system contractor and one main process equipment contractor only.

The Project

The goal was the realization of a fully automated 14,000-square-meter full-scale manufacturing facility for a drug that was produced by using recombinant DNA technology. A fast-track approach was taken for the project execution, in which the target "18 months from detailed design to qualification" was met.

The Control System Project Scope

Turnkey delivery of process modules from selected main equipment suppliers for different process areas in the new plant covered everything from detailed design and construction up to commissioning and qualification. The automation was done by using a Distributed Control System (DCS) and information management system.

The main equipment suppliers prepared all Functional Specifications (FSs). The suppliers—together with NNE and the user—monitored the software implementation and executed all tests for the automated process equipment. The delivery included the preparation of all Installation Qualification (IQ) and Operational Qualification (OQ) protocols and their execution, following comprehensive requirements from the customer.

In order to fulfill the fast-track requirements, a model was developed for qualification. Qualification was executed partly in parallel with and (as much as possible) in combination with Factory Acceptance Tests (FAT) at multiple construction sites across Europe. In particular, it was mandatory to test and qualify all configuration and application software at the control system contractor test site.

System Overview

The installed control system was a standard DCS with an information management system based on available standard components and systems. The system handles the following functions:

- Recipe management
- Batch management

- Data collection

- Batch and genealogy reporting

- Process control and supervision

- Simple warehouse functions

- Manual material charge with raw material tracking information for logging

- Instrument and equipment maintenance

- Security system functions

- System network communication

- Communication with external devices

- Interface with equipment supplier control systems

These functions were implemented by using built-in functionality in the control system and by configuring the system and application software to achieve the required functionality. Multiple process equipment contractors were responsible for delivering completely tested and documented process modules with process equipment. These process modules contained the total system, including control systems with application software (see Fig. 5.1).

Front-end Definitions and Engineering

For large-scale pharmaceutical projects, it's mandatory to plan the process and automation engineering approach to cover all aspects of the life cycle, from user requirements through implementation to validation. By evaluating opportunities and business risks, the main equipment contractor for this project ended up with three prequalified and selected main equipment contractors for process packages:

- Media preparation and fermentation (Sartorius BBI, Germany)

- Purification (Millipore, France)

- Stock solutions and solvents (Semcon, Sweden)

One main contractor was in place for control systems (Emerson Process Management, UK). Several other constellations of contractors were in place for utilities (e.g., process support and common process systems) and building services.

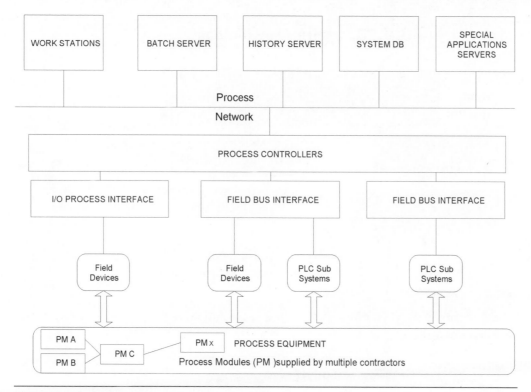

Figure 5.1.   Principal system structure diagram.

NNE owned the responsibility for global project management, construction management, engineering management, and quality management. The aforementioned equipment contractors and control system contractor agreed and were committed to the overall project execution model (see Fig. 5.2), where special attention was paid to the testing phase as well as for the split into process modules.

The overall basic principles for process and automation planning were the following:

- Get the user involved in the FS and design
- Agree on the recipes up front
- Do not implement until you have designed
- Understand the process before generating the design specifications
- Document the design

Front-end engineering definitions (decisions) followed strategic and conceptual activities and deliverables (from a control system perspective):

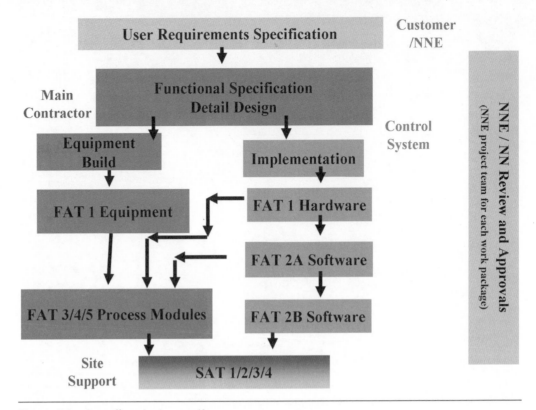

Figure 5.2. Overall project execution.

- Standards and guidelines
- Modularization principles
- User requirement specification
- The control and operability philosophy
- Specification tasks
- Contracting philosophy
- Master plan for validation and detailed validation plans
- Qualification tasks
- Quality plan
- Project life cycle activities

The position of the activities during front-end definitions and engineering is illustrated in Figure 5.3.

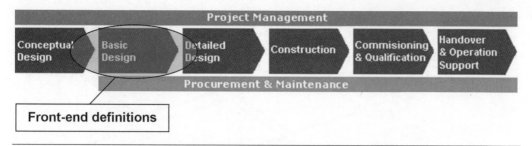

Figure 5.3.   NNE project activity model with position of front-end definitions.

The international standards and guidelines that were used as references for this project are the following:

- *ISPE GAMP: Guide for Validation of Automated Systems*
- *ISPE Baseline Pharmaceutical Engineering Guide, Volume 5: Commissioning and Qualification*
- *ISA-88.01-1995, Batch Control Part 1: Models and Terminology*

In order to achieve the fast-track project approach it was necessary to run paralleled activities such as building activities on site and process equipment (divided into process module entities) construction, testing, and qualification off site. These modularized methods were based on the NNE engineering solution for modularized plant facilities and manufacturing processes. These methods broke building and process down into building modules specifications and process modules specifications, where process modules consist of process equipment (i.e., process unit specifications) including the required control system for processing functionalities, qualification, and documentation (see Fig. 5.4).

User Requirement Specifications

The User Requirement Specifications (URS) were divided into two main categories (from the perspective of control systems). One part of the specification related to the Controlled Process part of the computerized system. This included process equipment to be controlled and operating procedures that defined the function of such equipment as well as manual operations that didn't require equipment (see Fig. 5.5, "Controlled Process" block).

Another part related to the control system hardware, system software, and applications software that controlled the operation of the computer (see Fig. 5.5, "Control Systems" block).

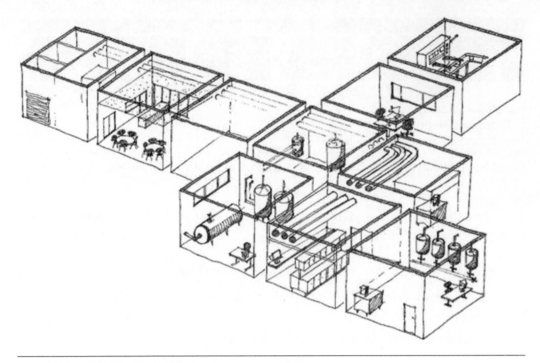

Figure 5.4. Principle for building and process modules.

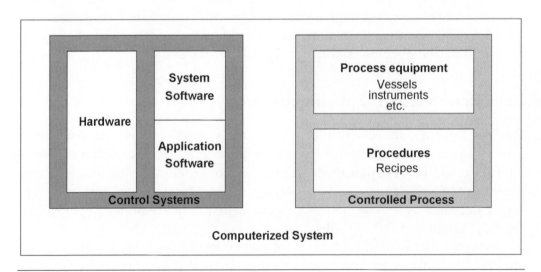

Figure 5.5. Definition of computerized system.

An overall requirement specification for plant control and operability philosophy was prepared by defining high-level functionalities that were control system independent. The plant description included the following items from ISA-88:

- A physical model
- Equipment entities and recipes
- Operational modes and states (with added hygienic states)
- Batch control activities

In addition, the following were also included:

- Clean In Place (CIP) and Sterilize In Place (SIP) methods
- Instrumentation
- Alarms, warnings, and messages
- Exception handling states
- Human Machine Interface (HMI)

All requirements were jointly created by the customer and NNE. Main contractors executed engineering activities, such as Process Flow Diagrams (PFDs), Piping and Instrumentation Diagrams (P&IDs), FSs, and so on, which described detailed "solution" responses to the requirements for each particular package.

Functional Specifications

When focusing on control systems as well as equipment functionality, the FSs are the key documents to creating a common baseline for functional understanding between customer, engineer, and contractors. Wise management opens this up for opportunities to reuse and make consistent aligned processing requirements between contractor packages for plant operability. Ideally the FS process should ensure the following:

- Customer users have the opportunity to make sure that their requirements are considered
- The FS is carried out in a rational, practical, uniform, and documented way
- Important design decisions are not made by programmers while coding

The FS process for this project (named Functional Requirements Analysis) included the following:

- The decomposition of the process and recipes into the ISA-88 physical and procedural objects to be represented on a set of diagrams and associated tables

- The detailed design of each of these objects is the basis for programmers to provide systems that meet the functional requirements

The resulting design was contained in an FS model. The methodology was based on the models and terminology defined in ISA-88 (IEC 61512). There are many interpretations of ISA-88. Different understandings of operations versus phases are common and the same applies to Equipment Modules (EMs) versus Control Modules (CMs). Unless the standard is further detailed and analyzed by company or project standards, there is still a lot of room for inconsistencies and misunderstanding.

To manage this, NNE issued a reference model (object database) for the process equipment contractors, "the FS model," which clarified the extent to which the choices that are allowed in ISA-88 apply to the project. It also showed how Units, EMs, Phases, and so on were defined. The reference model principle provided a sample of generic control mechanisms that were common to the entire process (such as transfer phases, common resources, and in particular, a set of agreed CMs). The model illustrated in general views how the diagrams and tables defined the required control and operational activities (in plain process function language) without control system influences. The model included the following:

- Generic control requirements for both normal operation and exception handling

- The structure of the model, based on ISA-88

- The use of diagrams

- The use of matrices (equipment states)

- The use of tables and lists of data (parameters)

- The way that various control objects (items such as units, EMs, CMs, common resource modules, unit procedures, etc.) were defined

- HMI aspects of the model

The model used defined two of the ISA-88 models and the relationships between them (see Fig. 5.6):

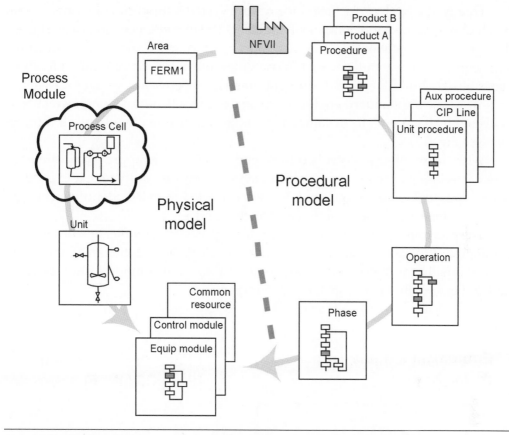

Figure 5.6. ISA-88 physical and procedural model used for FS development.

- *The physical model.* This includes the basic control functionality in terms of site, areas, process cells (with identification of Process Modules), units, EMs, and CMs.

- *The procedural model.* The recipe is defined in terms of procedures, unit procedures, operations, and phases.

As mentioned earlier there is still a lot of room for inconsistencies, misunderstanding, and plant-wide functional uniformities. To improve this, we executed "white smoke meetings"—workshops before starting major activity, which included the preparation of FSs and the commissioning and qualification phases. At these meetings, all practical issues and concepts were described. Solutions were decided and adopted by all parties. Further monthly global meetings were held for all contractors and NNE to sort out conceptual engineering and coordination issues, engineering status, and so on.

Finally, the NNE project team for control systems supported the project with two highly qualified and competent engineers (system engineer and overall practical process automation planning engineer) who coached the main contractors' engineering teams about proper interpretation of the reference model (Toolbox) about how to build the models and get uniformity between packages, as well as on how to use the standard engineering database tool (Control Draw), which was delivered by the project to all contractors and recommended for use by all contractors (see Fig. 5.7).

With the previously described front-end decisions and approved documents, a final baselining of scope for the control system, process equipment, and production recipes took place. Requirements were corrected to reflect identified changes to achieve realistic production process functionalities.

The previously defined structure was loyally used as identification of elements in the project for downstream engineering executions by the different contractors both internally for definition terminology, naming, and communication convention and externally to manage the following issues:

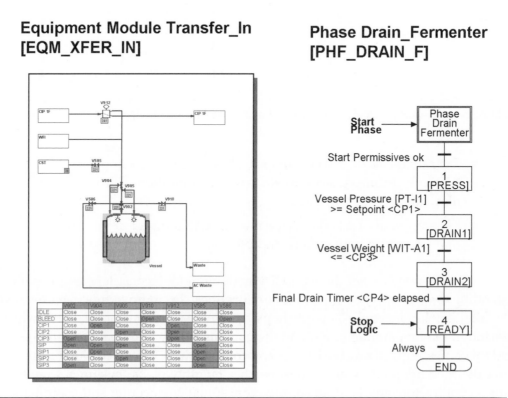

Figure 5.7. Typical drawings with data for physical and procedural models (database objects).

- Implementation
- Construction
- Schedule activities (erection, Commissioning and Qualification [C&Q] activities, etc.)
- Tests off site and on site
- Process module delivery lists
- Code reviews
- Module and integration tests
- Commissioning off site and on site
- Qualification off site and on site
- Technical documents

Engineering Execution Activities

Based on the scope and modular breakdown and naming definition detailed previously, the individual contractors mainly executed the deliveries according to their own shop floor workflow and manufacturing procedures for process modules. The control system contractor managed the control system according to the same definitions. (Again, the focus is the perspective for the control system.)

Detailed Design

With complete FS or agreed complete split (using both a physical and procedural model), the control system contractor prepared the Functional Design Specifications (FDS), which detailed the required functionality from the FS for the main equipment contractors by using direct data pasted from the FSs where applicable. Further design specifications were prepared for the control system if required for programming details.

Functional Design Specification and Hardware Design Specification

The FDS was developed from the URS and from the FS. This document described how the control system should operate and how it meets the complete system requirements. The FDS contained detailed information on the interfaces between the computerized system and its operating environment. The control system contractor produced a number of Detailed Design Specifications (DDS) for more specific information detail before programming. Lead engineers from the control system contractor were placed at the main contractor locations to implement and transform the FS required functionalities into FDSs. To ensure traceability of functionality, one FDS was provided for each FS. The hardware design specification defined the control system hardware for the system.

Test Activities

The overall main activities for testing (including qualification) were the following:

- Process equipment tests (physical properties of equipment)
- Process Control System (PCS) test (hardware, system software, applications software and network, etc.)
- Recipe test (execution of recipe on the equipment)

Acceptance testing was divided into factory (FAT) and site (SAT). The following section describes the main testing activities associated with the Process Equipment. The naming conventions FATx and SATy are "activity slicing" and are used to identify the testing activities in a manageable and practical way and support the opportunity to create consistency across packages planning for parallel testing activities. This allowed the flexibility for each process module to be tested off site or on site to a certain level. The naming of tests for "activity slicing" is in accordance with ISA-88 with other practical testing activities at that state of the particular module.

Process Equipment Testing Activities

The testing—including qualification for dedicated process areas—was split into several parts. The relevant parts of the testing are shaded in Figure 5.8. Non-shaded parts of the testing are specific to control system software and hardware. These test activities are detailed in the "Control System Module and Integration Testing and Qualification" section, later in this chapter.

The process equipment for different process areas was delivered to the site in complete functional process modules. This facilitates their qualification on a modular basis with system qualification testing to be executed, where possible, at the main contractor's site. Integrity testing of the installed process EMs on site was executed at the start of the on-site qualification activities (IQ/SAT1) to verify that none of the qualification activities executed prior to site installation were invalidated.

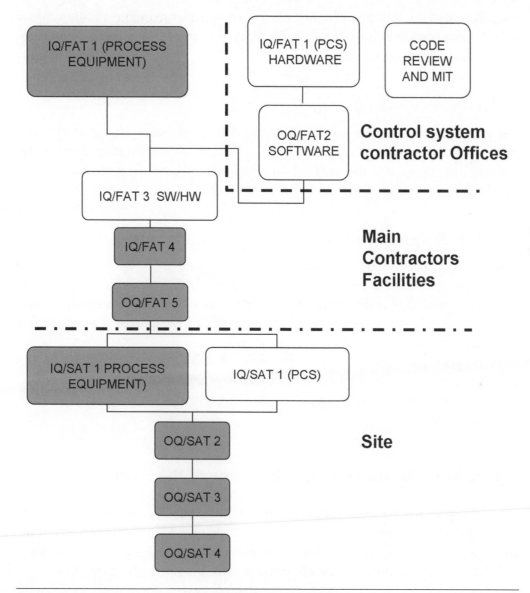

Figure 5.8. Structure of testing and qualification activities.

Installation Testing and Qualification at the Factory

IQ/FAT1 (Process Equipment) Test

The purpose of this testing was to verify and document that equipment, instruments, valves' piping, and other components of the system were present and installed correctly, as defined by the URS and design documentation, and to ensure that the installation adheres to all prevalent regulations and standards, prior to loop checking and functional testing.

IQ/FAT4 (Process Equipment) Test

The purpose of this testing was to verify and document that all instrumentation is installed and functioning correctly, as defined by the URS and design documentation, and to ensure that the installation adheres to all prevalent regulations and standards, prior to functional testing of the system.

IQ/SAT1 (Process Equipment) Test

The testing also verified and documented if all connections to the system were present and installed correctly, as defined by the URS and design documentation, and ensured that the installation adhered to all prevalent regulations and standards. This testing also verified the integrity of the process modules after installation on site.

This testing verified that the system was in the same condition as it was when it left the contractor shop floor and identified any areas that might have required retesting, including any retesting of control loops or retesting of process equipment functionality.

Operational Testing and Qualification at the Factory

OQ of the system is documented evidence that the system operates according to the URS and the FS through all anticipated ranges. It includes identification of all important operating parameters, anticipated ranges, and appropriate acceptance criteria, and the tests are performed to demonstrate that the system meets the criteria.

The system OQ was executed in four distinct parts located in the equipment contractor shop floor or on site. The OQ of the integrated system was executed according to the project defined delivery plan, before delivery of the process modules to site. The four different parts of the testing were as follows:

1. OQ/FAT5

- Testing of any equipment functionality (ISA-88 EMs and CMs and ISA-88 phases) that could be tested without the supporting plant utilities

2. OQ/SAT2 (Control system and process equipment)

- Testing of supporting utilities and connections and transfers between individual units and process modules

3. OQ/SAT3 (Control system and process equipment)

- Completion of all testing of equipment functionality, incorporating all ISA-88 EMs and phases that could not be completed as part of OQ/FAT5

4. OQ/SAT4 (Control system and process equipment)

- Testing of recipes and procedures

All commissioning and OQ tests for the process equipment and the recipes applied stringently to ISA-88 structure naming and the functionality tested according to the FS models. Figure 5.9 illustrates a split example of schedule applying to ISA-88 terminology. The same methodology was applied for planning and execution of commissioning and qualification tests. Process units and EMs are tested on all pieces of equipment on which they occur. Phases are basically only tested once on each type of equipment.

Control System Module and Integration Testing and Qualification

This section describes the main testing activities associated with the control system software. Qualification and testing of hardware are related in Figure 5.8.

Software Module and Integration Test Specification

The software module and integration test specification describes the structural tests used to verify that each software module met the requirements defined in the FDS and that the modules communicated properly with each other as defined in the FDS.

Module Testing

This is the process of evaluating application software modules based on their performance against design specifications. Module testing was performed on manageable segments of code with a level of complexity that allowed for the exercising of all module functions. Module testing must take into account the internal function of the software module.

OPERATIONAL TESTS TEST TYPE			Tank XXX	Tank YYY	Mixing X	Reactor A	Tank ZZZ
Test Type	Test	Test	302A	310A	120A	120B	120S
EQM	OQ	Pressure Control Test	▒	▒	▒	▒	
EQM	OQ	Agitation Control Test	▒	▒	▒	▒	
EQM	OQ	Aeration Control Test				▒	
EQM	OQ	Interlocks Test	▒	▒	▒	▒	
EQM	OQ	Pressure Hold Test	▒	▒	▒	▒	
PHASE	OQ	Full Vessel SIP Test	▒		▒	▒	
PHASE	OQ	Empty Vessel SIP Test	▒		▒		
EQM	OQ	Empty Vessel SIP Test					▒
EQM	OQ	SIP Test (Transfer Routes)		▒	▒		
PHASE	OQ	SIP Test (Transfer Routes)					
PHASE	OQ	Liquid Filters SIP Test			▒		
PHASE	OQ	Liquid Filters Integrity Test			▒		
EQM	OQ	WFI Flush Test (Routes)	▒			▒	
RECIPE	OQ	CIP Vessel Recipe Test (Full)			▒		
RECIPE	OQ	CIP Vessel Recipe Test (Reduced)					
RECIPE	OQ	CIP Line Recipe Test			▒		
RECIPE	OQ	CIP Filters Recipe Test			▒		

Figure 5.9. Typical OQ test schedule applying to ISA-88 structure as key index.

Module Integration Testing

Module integration testing is the process of testing combined software modules to evaluate the interaction between them. These tests examined the transfer of data and control parameters across the module interfaces and demonstrated that the desired functionality could be supported.

System Installation Testing and Qualification

The purpose of the IQ is to provide documented evidence that the delivered control system has been installed to system design specifications and that drawings and functions are in accordance with specifications. The testing is performed with the actual hardware and software that will be part of the installed system configuration. The testing is accomplished through either actual or simulated use of the software being tested within the environment in which it is intended to function.

The IQ was subdivided into three distinct steps for execution. Each part required a postexecution approval, which needed to be completed before proceeding to the next part in the qualification of the control system.

Step1: IQ/FAT1 Test

The purpose of this testing was to ensure and document that the hardware and system software with all its components were operating correctly as defined by the URS, hardware design specification, and DDS prior to application software testing.

Step 2: IQ/FAT3 Test

The IQ/FAT3 Test Specifications described the testing to be performed at the various equipment contractors' sites after the completion of OQ/FAT2. The purpose of this testing was to ensure and document that the hardware and software installed were the correct versions, and that all hardware had been installed per the manufacturers instructions, URS, and the drawings.

Step 3: IQ/SAT1 Test Specification

The IQ/SAT1 Test Specification describes the testing that was performed on the PCS hardware and software after the arrival of the equipment on site.

System Operational Testing and Qualification

OQ of the automation system is documented evidence that the system operates according to the URS and the FSs. OQs may be performed on the integrated system or on each subsystem. It includes identification of all important operating parameters, their anticipated ranges, appropriate acceptance criteria, and the tests that will be performed to demonstrate that the system meets the criteria.

The software OQ was executed at the control system contractor office as part of the documented OQ/FAT2. It was performed by the control system contractor and witnessed by representatives from the main process contractors. All control sequences were exercised using simulated Input/Output (I/O). Qualification of the control system integrated with the various process modules is detailed in equipment testing and qualification (as described previously).

The response of the control system to both normal and abnormal test cases was recorded and evaluated against predetermined acceptance criteria derived from the URS, FSs, FDSs, and DDS. All tests were based on predefined and documented parameter settings. Boundary tests were performed where applicable.

OQ/FAT2

Testing was split into two parts, FAT2A and FAT2B, for project timing reasons. The OQ/FAT2A covered similar aspects to the software module and integration testing. FAT2B was mainly concerned with the higher-level functionality of the overall system.

Change Control and Configuration Management

Change was controlled by the control system contractor from the start of qualification activities (FAT1). Configuration management was implemented by the control system contractor from the beginning of testing activities. The control system contractor produced a configuration management Standard Operation Procedure (SOP) for approval. This document detailed which items would be under control and how this control would be performed, documented, and audited. This document also specified how software versions were controlled during the acceptance testing at different main process equipment contractor sites and how storage and delivery were managed.

Project Management

This type of project provides real challenges for the project organization (including the contractor's project team) and must be based on global project management with commitment. The innovative modularized process breakdown set the project apart from conventional engineering and construction and required thinking in systems. NNE, as overall project managers, needed to know the details and keep the entire project process in sight at all times. Some key aspects of the project managment process (from a control system perspective) were the following:

- Management of several project teams
- Clearly specified roles and responsibilities
- Supervisory and advisory roles for contractor packages
- Management of interface points
- Management of the schedule and its critical path
- Clearly defined communication rules
- Awareness of contractual differences

Figure 5.10 illustrates the organizational relationships in the project.

Lessons Learned

The following are key issues for successful project execution:

- Always use a structured (ISA-88) approach and break the work down into manageable, reproducible elements. Follow standards and use consistent terms starting with structuring the URS splitting, if possible.

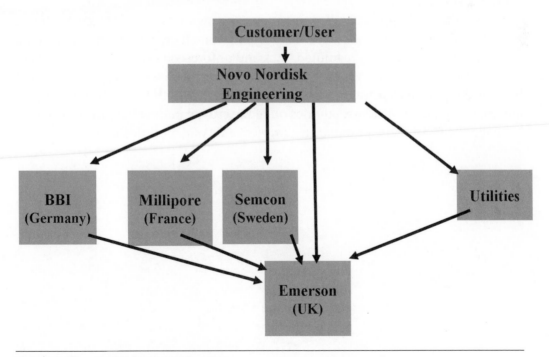

Figure 5.10.   Contractual relationships.

- Make a plan and stick to it.

- Plan wisely. Time spent at the front end of a complex project for definition and agreement facilitates easier implementation.

- Get the shareholders (users) involved from the beginning.

- URS and FS structures dictate the qualification tasks—put your effort into good specifications.

- Focus on the management of complex "distributed" project execution. This is the most demanding component and brings with it a higher risk factor.

- Never change "agenda" at handover.

Conclusion

This chapter described one control system contractor's use of ISA-88 in the project execution for a project with multiple stakeholders. In conclusion, we advise contractors to seek and create decoupling wherever possible; use it proactively for parallel-distributed activities and with the process module and ISA-88 structured project execution.

A staging test approach allows for early testing and a more effective utilization of a subcontractor's resources. This offloads on-site activities and opens up a more flexible schedule where limited slack and unforeseen events can be absorbed.

A strategic approach for the use of ISA-88 control elements typically used at biotech and pharmaceutical process equipment increases the opportunity to move automation (control system) activities off of the critical path of the project. (That is, by using the FS models for database libraries of ISA-88 modules.)

Using ISA-88 Batch Techniques to Manage and Control Continuous Processes

Presented at the WBF
North American Conference
April 13–16, 2003, by

David A. Chappell
Batch Technology Manager
chappell.da@pg.com
Procter & Gamble
8256 Union Centre Boulevard
West Chester, OH 45069, USA

Abstract

Have you ever thought of using batch sequencing and ISA-88 recipe management techniques to control a continuous process? This chapter will discuss the great similarities and subtle differences found in such applications. At Procter & Gamble we have successfully created several such "hybrid" process control applications. The results of these adventures greatly exceeded expectations. These systems benefited from an adaptation of the modular approach described in the ISA-88 part 1 standard with some significant differences in the states of their Equipment Modules (EMs) and Phases. The necessary modifications will be presented in detail.

Background

Continuous systems have existed alongside and in competition with batch systems throughout modern manufacturing history. Both approaches have their champions and both have advantages and disadvantages, which leads to competition between the manufacturing practices. Batch systems generally require less capital to build than continuous systems, but batch systems are considered more difficult to operate, which offsets the cost advantage. Batch is dominant in areas where material genealogy tracking is important, such as pharmaceuticals, with continuous dominating areas where mass production is important, such as petrochemicals.

At Procter & Gamble we apply a mix of these types of systems. Sometimes we have the same products made on both continuous and batch systems. There is continual debate as to which is the best manufacturing approach, and based upon business requirements, the right answer is sometimes both!

The examples I present in this chapter are similar to several systems that have been successfully implemented. Some of the aspects of these systems have been generalized for clarity, but all the significant details remain.

Batch Example

Figure 6.1 shows a simple batch system with two units and five different materials. In this example, the two batch units use weight scales to indicate the amount of material added. As such, they can only add materials sequentially, which will dictate the batch cycle time and therefore how much material can be created in a given period of time.

Continuous Example

The example continuous system shown in Figure 6.2 makes the same product as the batch example in Figure 6.1. It has all the same processing capabilities; they are just performed in a different manner. There is significantly more processing equipment involved in the continuous system than the batch system, making equipment capital costs typically higher for continuous systems than for batch. The continuous production is controlled by a final product flow meter against which all other flow meters are proportionally controlled.

Primary Physical Differences

The continuous system requires a method to independently monitor and control the amount of materials being transferred; in this example (Fig. 6.2), flow meters with control valves are indicated. While some batch systems will use this technique to improve cycle time, it is not mandatory and adds complexity and cost to the batch system.

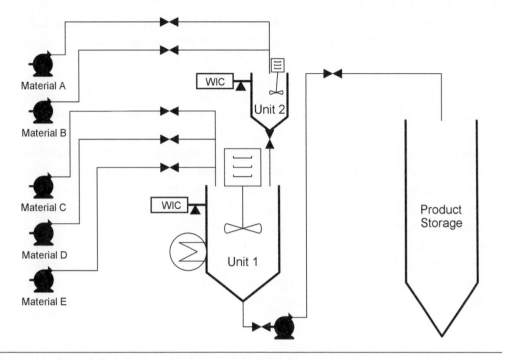

Figure 6.1. Simple batch example.

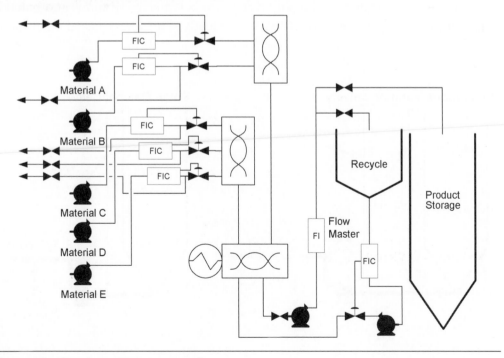

Figure 6.2. Simple continuous example.

There are six materials in this example because Recycle has been added. This is a common requirement because an indeterminate material is created during the transition times (e.g., startup, shutdown, and running product change) of the continuous system. Rather than waste this material, it is common to blend it back into the final product. The philosophy in the batch system is to make sure the product produced is right before sending it to product storage; therefore no recycle handling equipment is required.

Material handling can vary significantly between batch and continuous systems, as illustrated in Figure 6.3. In continuous systems, recirculation lines are generally included in the material delivery systems to reduce the amount of recycle produced. The recirculation lines provide a method to fill the supply line and guarantee proper operation of the flow monitoring instruments before the full system is started. In batch systems using weight control, this does not affect accuracy, since material delivery to the receiving unit will only be delayed by empty lines. In both the batch and continuous systems, when a material is not being used for a given product, the supply lines are cleared by using compressed gases.

Primary Operational Differences

I have heard some state that the only difference between continuous and batch is that continuous is a single batch with a long, steady state. I have heard others state

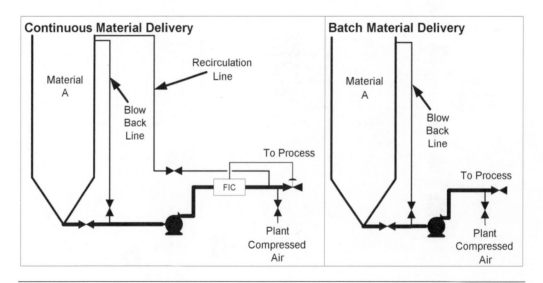

Figure 6.3. Material delivery.

that a batch is just a continuous system that is in a continual flux of startup and shutdown. And one good friend of mine says they are "exactly the same . . . except for what's different."

People who have worked with both know that "what's different" is significant. The focus of continuous systems is "running," which can be hours, days, weeks, or longer. Continuous systems can generally make more material in a shorter time than comparable batch systems; this is one of the primary advantages of continuous systems. The traditional focus of operations for continuous systems is the startup, which is a one-time event; these systems mostly take care of themselves after an operator has "lined out" the system, whereas batch systems have traditionally required "continuous" attention by operations as product is made. Thus we see a trade-off in the cost of equipment versus the cost of people.

Primary Automation Differences

The traditional approach to automation of a continuous system has been to rely on industry standard controllers that will allow the flows of materials to maintain a ratio against the total flow of finished product. This is well-understood technology and easy to successfully implement almost anywhere. If there is any automation in the startup of the system, it traditionally has been custom, leading to very complex monolithic systems that only a few people understand and even fewer can support and modify. As with older batch systems, any change in product formulation becomes difficult to manage.

Since the early 1990s, batch systems have received a lot of attention because of the ISA-88 standard and the many newly available products to automate the "continuous starting and stopping" of batch manufacturing. This has led to many products that are designed to manage both a "recipe" for a product and the "sequencing" of a process to create that product. The "modularization" techniques recommended by ISA-88 have proven to be extremely effective in the automation of batch systems, and they have become the batch industry standard approach. Many of these automated systems now run with even less attention than their continuous counterparts.

During the early days of the ISA-88 definition effort, it was observed that the standard could apply to all forms of manufacturing and not just batch! While most everyone agreed with this, Thomas Fisher had the wisdom to keep the focus only on batch. His point was that there never would be an ISA-88 standard if we attempted to reach too far, and continuous was too far at that time and could still be today, but it may not be tomorrow.

A Merging of Technologies

In three successful applications, P&G has been able to blend batch automation and continuous automation technologies to create systems where we automate management of product formulation and recipe management using "batch" technology, automate the startup of a continuous system using the "sequencing" capability of "batch" technology, and retain the traditional continuous operation automation that is so comfortable to manufacturing.

This has required a change in mindset "modularization" of the continuous systems by those responsible for the process and its automation. This is no small task, but once done the benefits are eventually obvious and it becomes the property and philosophy of the engineers who own these continuous systems. It is unfortunate that they do not have a method to broadly share this information, since it seems to reside in small groups and is not widely accepted.

The implementations consist of three independent "managers": one for recipe management, one for the startup sequencer, and one for continuous operations, as shown in Figure 6.6.

Recipe Management

The recipe manager uses the standard ISA-88 guidelines and is a commercially available package, which we are able to use "as is." This has allowed all those in charge of the product specification to have their intent directly applied to the actual production, and this has had a significant positive impact on data entry mistakes.

Startup Sequencer

The startup sequencer uses a standard ISA-88 Equipment Phase to interface with a commercial recipe sequencer, which manages three phase types: initialization, material, and sequencer run. The phase states and actions are described in Figure 6.4.

Each phase interacts with components of the continuous manager. The Initialization phase clears all pending material information and informs the continuous manager that a new recipe download is in progress. The Material phase provides the materials that are used in this recipe with all the information necessary to be used and issues a request for those materials to proceed to their ready states, if they are not already there. The Sequencer Run phase provides all other recipe processing information, such as temperature and residency times, and informs the continuous manager that all materials are ready for production; it then terminates

Sequencer Phase States

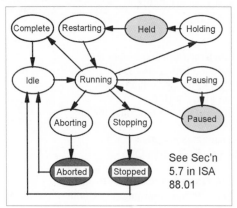

See Sec'n 5.7 in ISA 88.01

Sequencer Actions

Interaction of Sequencer Initialize Phase and
Continuous Material Equipment Module
    Clear all Sequencer Material Information
    Complete Sequencer Initialize Phase

Interaction of Sequencer Material Phase and
Continuous Material Equipment Module
    Provide Material Information
        If not already Running Request a Start
            Wait for flow indication
    Complete Sequencer Material Phase

Interaction of Sequencer Run Phase and
Continuous Manager
    Inform Continuous Manager all materials are
    ready for production
    Complete

Figure 6.4.   Sequencer phase states.

the recipe startup sequencer. Functionally, the startup sequencer replaces the manual operation of bringing the material paths to a useable state. It eliminates the varying effects of different operator approaches to this, resulting in much smother operation and better quality product sooner.

Material EMs

The physical equipment represented in Figure 6.3 uses the ISA-88 Control Module (CM) concept to manage each individual piece of equipment: one for the pump, one for each block valve, and so on. These CMs are in turn controlled and coordinated by an EM that orchestrates the CMs to carry out the necessary process functions. The Material EM states and actions are represented in Figure 6.5.

The Material EMs provide a "bridge" between the startup sequencer and the Continuous Production Manager (CPM), as shown in Figure 6.6. As indicated, the Material EM can receive transition commands from three sources: the startup sequencer, the CPM (by operator action), and itself. The startup sequencer provides recipe information in a "pending" storage area for the Material EMs; upon production startup, this information is transferred to a "running" storage area and will be used for production. The operator has access to all this data and can modify it within the parameters of the recipe. By creating a single modular EM for material transfers and then reapplying it for all material transfers, the engineering effort is greatly reduced over traditional approaches. It also proves to be much

Continuous Material Equipment Module States and Actions

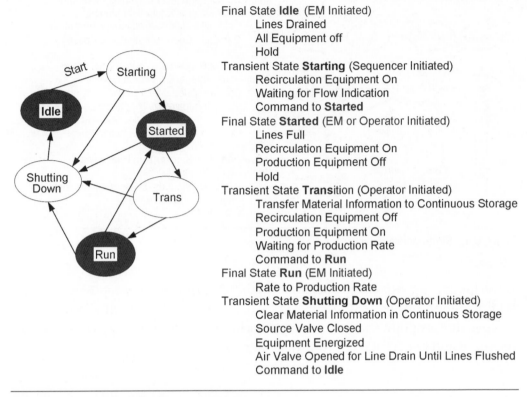

Final State **Idle** (EM Initiated)
    Lines Drained
    All Equipment off
    Hold
Transient State **Starting** (Sequencer Initiated)
    Recirculation Equipment On
    Waiting for Flow Indication
    Command to **Started**
Final State **Started** (EM or Operator Initiated)
    Lines Full
    Recirculation Equipment On
    Production Equipment Off
    Hold
Transient State **Trans**ition (Operator Initiated)
    Transfer Material Information to Continuous Storage
    Recirculation Equipment Off
    Production Equipment On
    Waiting for Production Rate
    Command to **Run**
Final State **Run** (EM Initiated)
    Rate to Production Rate
Transient State **Shutting Down** (Operator Initiated)
    Clear Material Information in Continuous Storage
    Source Valve Closed
    Equipment Energized
    Air Valve Opened for Line Drain Until Lines Flushed
    Command to **Idle**

Figure 6.5. Material EM states and actions.

easier to support than previous monolithic systems, and necessary changes to the process are now more easily accomplished.

Continuous Production Operations

The CPM has three modes: "Full Auto," in which it will rely upon the Sequencer for "advice" and the Operator for confirmation; "Semi-Auto," in which the operator can assume direct control of the EMs, bypassing the Sequencer and CPM but taking full advantage of the EM logic; and "Manual Mode," in which the operator takes direct control of all CMs bypassing all other control and logic.

In Full Auto during the initial startup of a recipe, the CPM is waiting for indication that all required materials have reached the "Started" state. Once that has occurred and the operator signifies the start of production, the CPM instructs

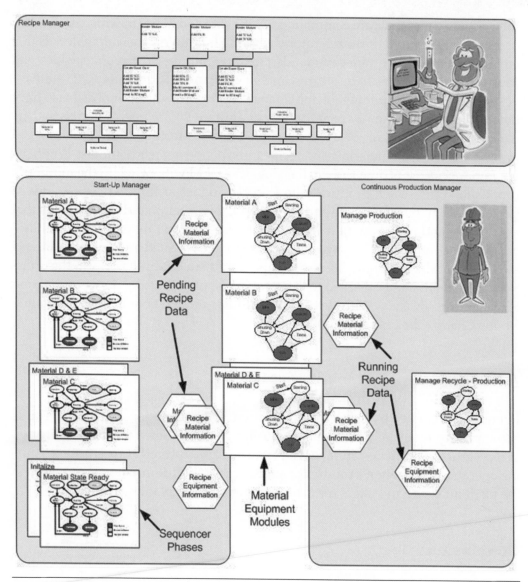

Figure 6.6. The total system.

all required materials to proceed to the Run state. During this time the production "Rate" is maintained at the minimum setting, and all material produced is directed to the recycle vessel. When the operator is satisfied that the product quality is at target, production is redirected to product storage and the production rate is set as desired. Any system upset affecting quality will cause the rate to be set to minimum and will redirect product to the recycle vessel, waiting for the operator's decision to hold, shutdown, or resume production. If the operator chooses "Hold," then all active materials are commanded to the "Started" state in which no product or recycle is produced. If the operator chooses "Shutdown," then all materials that are not already in "Idle" are commanded to the "Shutting Down" state, which will cause all material paths to be cleared and all equipment to go to its "Off" state.

There is also a "Running Recipe Change" option that makes it possible to have the startup sequencer load a recipe while running another recipe. Materials that are not part of the current recipe are commanded to the "Starting" state, and the CPM is informed when all materials and recipe information are ready for use. When the operator decides it is time to switch recipes, the CPM will switch to the minimum production rate, switch production to Recycle, and issue "Started" commands to all materials in the new recipe and "Shutting Down" commands to materials that are not idle and not in the new recipe. When the operator determines conditions are right, the command to switch to Product Storage is issued and the rate adjusted as needed. In the past, only a few operators could successfully perform this function, and even then, it often led to catastrophic results, shutting down production for significant periods. With the ISA-88 modular implementation, the "Running Recipe Change" has never failed!

Reporting

Reporting is handled by the CPM and manufacturing operations personnel. This is accomplished by using flow totalizers during the times when material is directed to recycle or to production storage. The totalizers are captured at specific times during operations and when a product change occurs. This information is provided to the plant information system for use by operations and plant management.

Conclusion

There are times and situations where continuous systems are appropriate and where batch systems are appropriate, and there are situations where either could be used. The use of the modularization recommended by the ISA-88 standard has significantly reduced the effort and cost of the automation of batch systems. While these modularization approaches have been demonstrated to improve the automation efforts of continuous systems as they have for batch, having this modularization approach accepted by those who do the automating is difficult, with most still following the traditional monolithic automation approach. Using the recipe management and sequencing capabilities of commercially available batch products has also greatly improved the operational management and execution of continuous manufacturing, although again, gaining acceptance is difficult. There are groups that believe strongly in the superiority of continuous manufacturing, and they will go to great lengths to prove it, including rejecting any automation techniques that are shared with batch systems. When the reluctance to use the ISA-88 concepts is overcome, continuous systems will realize the same improvements that batch systems have realized.

Batch and continuous are exactly the same . . . except for the differences.

Using ISA-88 to Automate Procedures at Continuous Processing Facilities

Presented at the WBF
Make2Profit Conference,
May 24–26, 2010, by

Dr. Wayne Gaafar
Batch Consultant
wayne.gaafar@honeywell.com
Honeywell International Inc.
Honeywell Process Solutions
1860 Rose Garden Lane
Phoenix, AZ 85027, USA

Bruce Kane
Batch Consultant
bruce.kane@honeywell.com
Honeywell International, Inc.
Honeywell Process Solutions
3079 Premiere Parkway
Suite 100
Duluth, GA 30097, USA

Abstract

All continuous production facilities have procedures that are only run periodically. These procedures include equipment startup, transition between product grades, switching between primary and backup equipment, and equipment shutdowns. These procedures can be very simple, in the case of switching between standby and running pumps, or incredibly complex, in the case of a cold plant startup. For the case of simple procedures, operations trained personnel to follow prescribed guidelines to ensure safety and timeliness. With more complex tasks, aids such as checklists and Standard Operating Procedures (SOPs) were developed to assist training. Producers have experienced that the more complex the procedure, the more variability in the quality and the timeliness of the procedure's results. Additionally, manufacturers can document the intellectual property that has historically resided in the minds of the operators for complex procedures. For multi-sited manufacturers this can increase the shared learning and reduce the risk of information loss as workers retire or leave their current position.

By using ISA-88 structures when automating complex procedures, manufacturers can see significant increases in procedure quality and reduced time to completion. This can lead to significant returns. We will show how procedures can be automated using ISA-88 structures. We will also show how the resulting procedures can significantly increase quality and reduce the time to completion. We will also show how these techniques capture and retain the intellectual property of these periodic procedures.

Introduction

Do the nomenclature, principles, and practices of ISA-88 (ISA-S88.01-1995) and batch processes have a home in a continuous process facility? If one only considers the long, steady-state, continuous nature of these processes, perhaps not. Manufacturers such as refiners would often cringe if their processes were compared to batch processes. However, every manufacturing process has a startup and shutdown, and almost every manufacturing process has transient states if one was to examine these continuous processes closely, as is done in Table 7.1. It is the long runtime of these so-called continuous processes that allow them to live up to their namesake, as they are mainly continuous in operation.

However, with thorough examination and study of processes in a continuous processing facility, a manufacturer comes to understand that not every process run in their facility is a purely continuous process. This is because many of those processes have very short time spans or are executed on a somewhat infrequent basis. These temporary states can include equipment startups, transitions between product grades, switching between primary and backup equipment, and equipment shutdowns. Any repeated series of steps to transition from process state A to process state B in a continuous process can benefit from the nomenclature, principles, and practices of batch.

Table 7.1. Example batch sequences in continuous processing

	Power generation	Polymer production	MMM refining	Olefin	Waste treatment	Utilities	Specialty chemical	Refining
Startup	X	X	X	X	X	X	X	X
Shutdown	X	X	X	X	X	X	X	X
Equipment Startup			X			X	X	
Equipment Shutdown						X	X	
Equipment out of Service	X	X	X	X	X	X	X	X
Reduced load	X					X		
Grade change		X		X			X	X
Product change		X		X			X	X
Material change		X		X			X	X
Unloading		X	X	X	X		X	X
Filling		X	X	X	X		X	X
Filter change		X					X	X
Die change		X					X	X

These process state changes can be relatively simple, such as switching between a standby and a running pump, or they can be much more complex, such as starting up a multi-column distillation area. In the case of these simple procedures, operations personnel are often trained to follow prescribed guidelines to ensure safety and effectiveness. The task is different when dealing with incredibly complex procedures. The safety implications alone require operators to take great care in engaging within complex areas. More complex procedures must be followed to protect both personnel and equipment in an area that may have been down for an unknown period of time. With increased process complexity there comes a greater potential for increased product variability. When complex procedures are executed manually, their impact on equipment, quality, and production throughput can be costly.

It is in these transient areas that batch principles should find roles to play in the otherwise uninterrupted world of continuous procedures. Specifically, an effective strategy for developing and managing these complex procedures is to employ ISA-88 automation standards, and in the future ISA-106, as it develops, will be employed to develop these procedures. For batch processes, ISA-88 principles have helped efficiently segment processes and procedures. ISA-88 has provided manufacturers with a common language and set of principles to describe their processes and, more importantly, improve their processes. Great opportunities exist for applying these same practices to continuous processes.

Data versus Legend

When one thinks of a continuous process, the mindset is to get the process to a steady state and "let it flow." Whenever a repeatable change is made, however, the procedures that transition the process between various states can be very intricate. For example, starting up an idle plant can prove very complex if the overall level of integration is high. Because integrated subsystems interact with each other, the procedure often isn't as simple as starting units in a specific order. In some respects, the startup is analogous to lacing-up a boot—each unit has to come up a little bit at a time before the whole area is up and running.

Currently the most common solution involves utilizing SOP documents that are designed to walk operators through these complex procedures step by step. These documents are generally based on methods that have developed over time. However, these documents are not always well written, complete, or even in some cases accurate, which puts an inexperienced operator at a significant disadvantage. Smaller, critical details that a veteran operator might know instinctively won't be as apparent to a less-experienced counterpart. As a result, some form of human intervention is often needed and "best-operator" problems arise. That is, plants may choose to execute critical and complex procedures only if their best operators are present. This often is due to the risk the manufacturer takes when running one of these procedures. Manufacturers should be concerned with the "intellectual property" that is held by these operators and not documented in the SOPs. This is the primary reason why knowledge retention is one of the most talked-about issues that will affect the future of the processing industries, be they batch or continuous.

This all contributes to an already risk-averse culture where plant decisions are moving from being "legend based" to data based. In essence, plant personnel are unwilling to alter procedures because, put simply, "it's always been done that way, and it works" or due to fears about the exact nature of the processes. By applying ISA-88 principles, these manufacturers can analyze and improve their processes by looking at those complex procedures from a batch perspective.

Let us consider an example scenario for the startup of a separations area. In many continuous plants, a series of interconnected distillation columns provide the separation of key components from a process stream. A simplified three-column train is shown in Figure 7.1. The steps involved in bringing the unit from a cold state involve the preparation of a column to isolate it from the atmosphere, the addition of a small amount of raw material, the initial setting of the correct temperature profile, the introduction of an input raw material, and finally, setting the output of products. Each column must undergo each of these operations. For one scenario for the startup of these columns, see Figure 7.2.

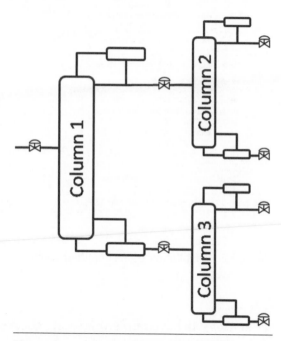

Figure 7.1. Simplified distillation train.

This is a common style of sequence that develops over time: either one person is responsible for the startup, or the unit starts as a single column and then other columns are added over time. In either case, the startup is done in a very serial manner.

This simplified example lacks the detail necessary to start an actual column. Each step could have a number of more specific steps. Additionally, actual

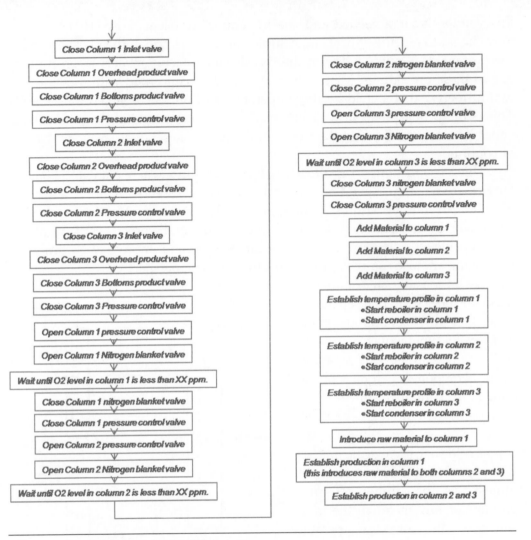

Figure 7.2. Distillation train simplified startup.

sequences will have numerous safety checks and reviews, as traditionally the components in these columns can be toxic, explosive, or both.

One nice feature of the serial operation is that it is very easy to understand. It can also be displayed on a clipboard as a series of steps. An operator can take the clipboard and check each step as it is completed. When the list is complete, the operation is complete.

When viewed in a block diagram, each step in this sequence appears to be a single block. This makes the sequence look like the process shown in Figure 7.2.

ISA-88 Illuminates: Looking at Continuous Processes in a Different Light

As noted before, most continuous process operators generally shy away from anything that directly compares their processes to batch processes. Therefore the first step in a successful ISA-88 application to a continuous process is to dispel the notion that ISA-88 is only designed for batch operations. Rather, ISA-88 should be viewed as a series of best practices that can be applied across industries to reduce procedure variability and improve overall safety, reliability, efficiency, and sustainability.

ISA-88's strengths are providing structure, a common language, and tools to devolve sequencing problems and capture process knowledge in a common framework. Using ISA-88, manufacturers can look at their continuous plants as batch plants with very long, stable points in their processes (the continuous sections). From an ISA-88 perspective, the start-to-end process could be viewed as one long batch sequence. Taking the normal, repeatable, and sequential parts of the process and applying ISA-88 principles, one can automate these procedures to make them more retainable, more documentable, faster, and easier for even inexperienced operators. The result is that the procedures are executed with the same level of quality consistently, regardless of operator experience.

Another benefit to this approach is the ability to directly address the ever-growing issue of an aging workforce in the process industries. Retiring veteran operators have a great wealth of knowledge about how plants are run. Due to poor intellectual property retention practices such as the use of outdated SOP books, this knowledge is lost when the veterans retire, and an inexperienced operator is left to fill the void. By using ISA-88 principles to document these operations, manufacturers can capture more of their intellectual property and include it in sequences that can be automated.

Applying the ISA-88 practices to continuous processing facilities entails dissecting each process, breaking it down into separate units, and then applying the appropriate amount of automation to each level. Applying these ideas to our simple distillation area, we could start by identifying each column as a unit. Then we could separate each portion of the sequence for the units into unit procedures. This would allow us to begin to group the startup sequence, as depicted in Figure 7.3. The next step would then be to identify operations on each unit. This could result in the startup depicted in Figure 7.4.

The steps under each operation can be said to be procedures. Because there needs to be some synchronization between the starting of each column, we can now separate the operations into phases with a synchronization point (Fig. 7.5).

Figure 7.3. Distillation train unit procedure identified.

Although what we have done so far appears to be just a renaming exercise, we can now recognize that the unit operations can be done independently. This allows us to create a startup sequence that allows parts of columns 1 through 3 to be started up in parallel, thus reducing overall processing time.

In practice, something as common as a distillation train is already recognized as a place where parallel activities can be performed. The application of ISA-88 techniques can, in more complex cases, identify time savings that would not be readily apparent.

This application of ISA-88 principles also enables historical engineering analysis that can allow manufacturers to discover problem areas and bottlenecks that have existed in their previous procedures. This is especially crucial because manufacturers increasingly need to examine every aspect of their processes as their companies continue to be squeezed by economic conditions. They must ask themselves questions such as, How can we make this process faster and better to benefit the bottom line? Time spent shutting down or starting up units or transitioning

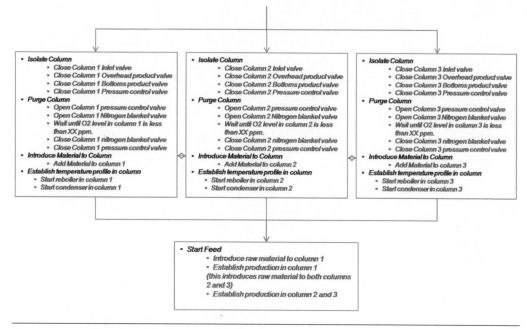

Figure 7.4.   Distillation train operations identified.

from one product grade to another equates to lost production—which means less profit. Therefore any time saved during these procedures is also money saved.

The actual realized savings will vary from industry to industry and plant to plant; if an existing SOP was well written and takes into account steps that can be taken in parallel, the savings generated by automation may not be as significant. However, if the procedures have largely remained unchanged throughout the decades, hours and even days could potentially be saved. In some cases, plants can reduce startup time by up to 20%, which amounts to additional profit. Even if procedures are executed only on occasion (e.g., once every 6 months), the effort is still worthwhile to eliminate extra time because the savings will still contribute to the bottom line.

ISA-88 Enlightens: Standardizing Engineering Analysis

Continuous plants typically do not conduct any type of optimization or analysis of their transient states. The closest measurement they employ is gross timing—that is, they know approximately how long it takes the procedure to run its course from start to finish but not much else. Additionally, the steps and functions performed during those transient states are often not well documented. This inability

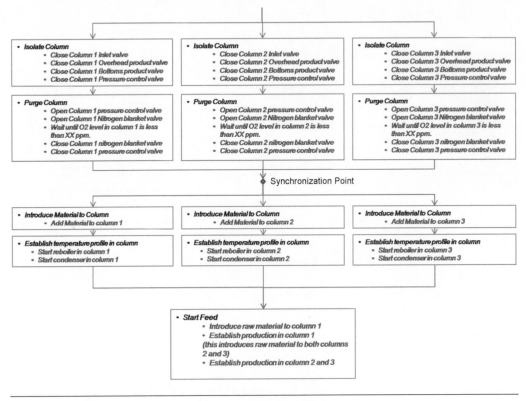

Figure 7.5. Distillation train sequence with ISA-88 applied.

to measure leaves little to no room for process improvement and makes it more difficult for plants to enhance overall operations to keep pace with increasing production demands.

Capturing sequences in a common ISA-88 framework enables engineering analysis that allows manufacturers to optimize sequences in current and other areas, as well as across global enterprises. This is largely because the ISA-88 principles examine the process from a modular perspective. By examining the modular components in similar units, experts can more easily critique and optimize sequences, point out omissions, and apply learning from other sequences. In this way, plants can uncover potential timing bottlenecks that could be causing one unit to start faster or slower than another. Once these problems are diagnosed, plants can take corrective action and implement best practices from common production sites to make the sequences safer, faster, and more reliable. The result is the establishment of business-wide "best practices."

As an example, take the distillation train: In this case we see three columns that have very similar procedures for startup. As these units startup, we can now keep detailed information about each operation and phase. This information can be items such as timing for each step or a calculation of energy used per pound of production during startup. By examining these variables between startups or units, variability can be identified and actions can be taken to reduce the variance. Because we applied the ISA-88 principles to the procedures in like processes, we are able to make these comparisons.

The ability to predict the duration of the procedures is a benefit of the ISA-88 application. Once the cycle-time variability has been reduced, the plants can then predict how long a startup will take and be able to plan production more accurately. Many manufacturers would rather predict when their units will be ready for production as opposed to quickly having a few units on-line while others lag behind. As the plants are brought on-line in a more deterministic fashion, the ability to coordinate the production logistics is improved.

The plants will also benefit by documenting the intellectual property that has typically been kept in the operator's head. In the past, this intellectual property was all too often lost whenever an operator retired or left the company. By applying ISA-88 principles, the manufacturers can capture this information and retain it even after personnel changes. As techniques for each type of process are adopted by a manufacturer, operators can be trained on these best practices throughout the manufacturer's facilities.

Conclusion

Whether continuous plants like to admit it or not, they have areas that are very much like batch processes. These processes can be highly complex and involved. ISA-88 provides a framework to analyze complex batch sequences.

In this chapter, a simplified example of a startup sequence was used to illustrate how ISA-88 can be applied. The application of these engineering principles was first able to aggregate a large set of steps into a series of unit procedures. Continuing to apply the tools to the unit procedures was separated into a series of unit operation and finally phases. Upon analysis, these unit procedures were seen to be independent. Thus a parallel structure for these procedures became the obvious choice. This parallel structure could possibly help reduce the cycle time for a startup.

The ISA-88 tools can be applied to the complex processes of continuous plants. When these tools are used, continuous process manufacturers can see benefits such as increased efficiency while running their processes and increased long-term support of their facilities. The cycle time of running these processes can be dramatically improved if parallel sequences are identified. Intellectual property can be retained by documenting the resulting procedures. This allows the manufacturer to retain their process knowledge and improve the efficiency of its workforce. Once the variability of these processes has been addressed, they become more predictable, which leads to better plant efficiency.

Do batch principles and practices have a home at a continuous process facility? The obvious answer now is YES.

Use of ISA-88 Techniques in Semi-continuous Applications

Presented at the WBF
North American Conference,
March 24–26, 2008, by

Dr. Wayne Gaafar
Consultant
wayne.gaafar@honeywell.com
Honeywell Process Solutions
2500 West Union Hills Drive
Phoenix, AZ 85027, USA

Abstract

In most continuous processing plants, there is at least one process area that acts as a semi-continuous process. Whether it is a water treatment facility, a set of carbon filters, or a set of reactors that foul over time, these processes all use sequences to switch units so that the set acts continuously. Setting up control systems designed for continuous operations to work for semi-continuous or batch operations can be difficult to program and maintain later on. By using ISA-88 batch conventions, semi-continuous processes can be easily implemented and maintained in modern controllers. A simple three-column semi-continuous operation will be used to demonstrate how the ISA-88 techniques can be applied.

Introduction

When developing batch applications, a set of standards have been codified that put a solid framework around the application. This ISA-88 framework allows the batch application developer to quickly and efficiently develop their application using techniques that have been proven very effective for batch processes. These techniques include compartmentalization of the various steps in a batch, the progressive disclosure of the actions that each step must take, the organization of actions both by physical equipment and by process flow, and a standard set of terminology that allows for easy communication between developers. When applied to a batch process, ISA-88 allows for quicker, more efficient development as well as easier modification and maintenance over the application's life cycle.

In continuous plants, batch applications are often not discussed or even considered. It is rare for the staff at a continuous plant to have knowledge of ISA-88 techniques or other batch conventions. Continuous plants tend to use terms like "sequence" or "manual procedures" to indicate operations that require timely intervention or a repeated sequence of steps. Whether they like to admit it or not, these plants do have batch processes, and they can gain from using ISA-88 structures to develop, run, and maintain their applications.

There are a number of processes that the continuous world calls semi-continuous and the batch world calls continuous batch. These processes are characterized by units that take a continuous input and deliver a continuous output but inside have a sequenced set of operations (a procedure). From an external view, these processes appear to be continuous, but inside there are usually multiple identical units running batch-like procedures. These similar units typically have discrete states with names like On-line, Off-line, Recovery, Regeneration, Ready, and Drying. These units tend to have one recipe that is continually run, effectively making the same product over and over again. Using either some process capacitance or overlapping on-line operations, the final product leaves the units in a continuous stream with little or no change in actual flow. Examples of these processes include the following:

- Water treatment using ion exchange resins or carbon filters
- Hydrogen purification using pressure swing adsorption
- Removal of water in a hydrocarbon stream using resin beds
- Lead or guard reactors before catalyst reactions
- Polymer reactions using multiple reactors in parallel
- Furnaces that utilize a de-coking operation
- Older refining reformers (before continuous catalyst regeneration)

Usually there are three or more identical units that are on-line or processing At least one of the units is Ready, and the other units are in various stages of

Regeneration or Preparation. When the On-line unit is nearly spent, one of the Ready units is switched on, and the spent unit is taken off-line and its preparation begins. This cycle repeats indefinitely and the inlet and outlet streams continuously flow.

Typical History of Semi-continuous Automation

If we examine the life cycle of a semi-continuous process, it typically begins as manual procedures where an operator follows a prescribed sequence of steps (using a printed set of instructions) that brings the ready unit on-line. Next, the operator takes the previous on-line unit and starts its regeneration process. Simultaneously, this operator performs the next procedure for the other units to eventually prepare them to go on-line. These processes are prone to errors and only occur at the speed of a human. As the inherent problems with a manual process are discovered (e.g., not every step performed the same, time variability between similar sequences, differences in off-normal handling, and different interpretations of the written step), Operations begins to explore alternatives.

Operations starts this by using processes and procedures that could be built into the continuous operations of the control system. These include alarm changes, the construction of interlocks, the use of simple logic blocks and combinations of interlocks, and blocks to automate the processes. Depending on the level of criticality of these processes, the plant might stop here and continue to do most of the task manually. When critical processes are examined further, automation is often pursued.

Automating these procedures introduces a new set of problems—namely support! As the complexity of the procedures increases, the number of people qualified to support them decreases.

With the introduction of structured text languages (e.g., FORTRAN, C, CL, Visual Basic) into the control system, many applications were written to solve these semi-continuous problems. Initially, these solutions did well for the straight-line scenario. (There were no problems with the process and everything worked as planned.) The code's ability to open and close valves, change setpoints, and to a limited extent, recognize the current state greatly improved the consistent running of these plants. These types of solutions all suffered from an inherent set of problems: the applications either ended up being too simplistic or the code became increasingly more complex as the number of "off-normal" conditions were recognized and addressed. The applications usually ended up being the bastion of one person in the plant. The phrase "Don't touch that—only Jim knows how to change that!" became all too familiar to the plant. The end result is that the applications were very static and rarely changed as demands on the plant changed. These problems can be minimized by using ISA-88 as an engineering framework for implementing automated procedures.

Sample Problem

To show how ISA-88 can be utilized, we will use a simple three-column model, as shown in Figure 8.1.

In the scenario presented in Figure 8.2 (a–e), the process stream flows through one of three columns where an impurity is removed. The amount of impurity is tracked until it reaches a predetermined level, at which point the "next" column is brought on-line (Fig. 8.2 b) and the current on-line column is switched from on-line to "regen" (Fig. 8.2 c-d) and begins a regeneration process. The regeneration process is simply another stream (labeled "Purge" in Fig. 8.2) that is fed counter to the process flow direction (Fig. 8.2 e) that removes the impurity from the "dirty" column until the column is ready to be placed back on-line.

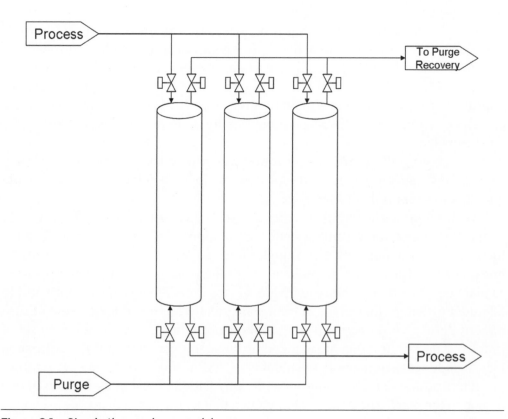

Figure 8.1. Simple three-column model.

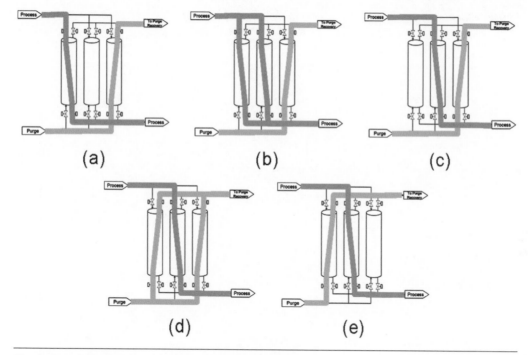

(a)   (b)   (c)

(d)   (e)

Figure 8.2. Three column model: Switching sequence.

Using ISA-88 to Solve the Semi-continuous Problem

We will now solve this example scenario using ISA-88 as a framework. The first step is to understand this framework. Figure 8.3 shows the basic building blocks in an ISA-88 model. One of the big advantages in using ISA-88 is that it is a well-thought-out methodology. (The developer does not have to reinvent the decomposition of their problem.)

We will start by defining the three columns as a Process Cell (Fig. 8.4); since each of the columns acts independently, we will let them coordinate their actions and thereby not have an overarching cell recipe. If the purity requirements changed from day to day or run to run, we might have made a different decision here.

Next we will look at the various operations that each column must run (Fig. 8.5). For this problem, we have divided the steps into the following states:

- *Initialization.* This state assures us that the column is ready to be placed on-line. That includes acquiring all the necessary equipment (e.g., valves, totalizers), placing all valves in a known state (closed), and resetting the total level of impurity.

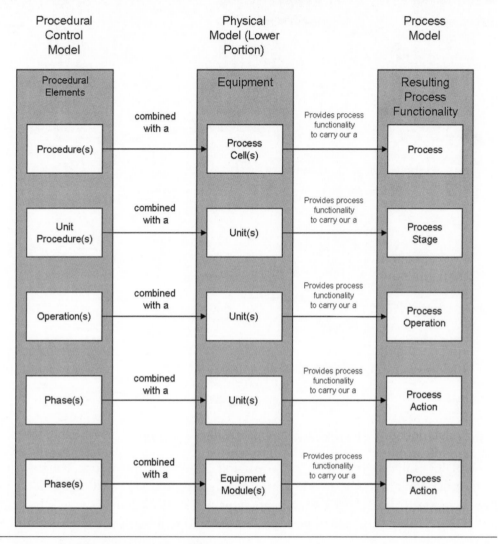

Figure 8.3. Building blocks of ISA-88.01.

- *On-line*. This state aligns the process valves to allow process to flow through the column. This state also starts the totalizer to tally the amount of impurity retained in the column.

- *Start-Next*. This state chooses which column will be started next and then issues the command to start that column.

- *Regen*. This state starts the regeneration process.

- *Ready*. This state lets the other columns know that this column is ready to go on-line.

As we drill deeper into the configuration of the application, we next look at the individual steps (Fig 8.6). For example, let's decompose the on-line state. To go on-line, the following steps must be performed:

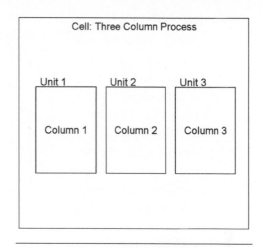

Figure 8.4. Cell and unit diagram.

1. *Acquire the valves.* This step is done to formally grab the equipment. This ensures that no other sequence or program will compete with this sequence for controlling this unit.

2. *Start the totalizer* (or request acquisition of the downstream impurity sensor). We need to count how much material is being deposited in the column. Then, knowing how much the column can hold, we can estimate when it is full.

3. *Open the outlet valve.* Steps 3 and 4 put the process flow through the column. Depending on process restrictions and interlocks, these steps could be reversed or done in parallel.

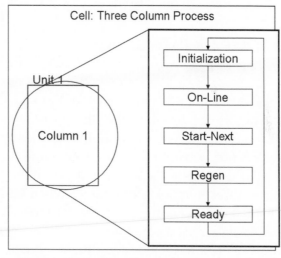

Figure 8.5. Column steps.

4. *Open the inlet valve.* See previous note.

5. *Check for process flow.* This step makes sure that everything is working. This check may be tried for a fixed period of time, after which if nothing is flowing then an alarm is sent to notify Operations that something is not right.

6. *Change state to "On-line."* Once everything is flowing, the state is put into a normal (on-line) state.

7. *Wait until the totalizer exceeds a set amount* (or until the downstream sensor exceeds a set value). When the totalizer reaches a predetermined amount (via the recipe), we begin the switching step. If we use a downstream sensor, then we will be watching the slope of the curve to identify an impurity breakthrough and also to switch.

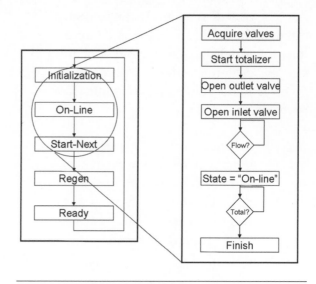

Figure 8.6. On-line state steps.

8. *Finish.* If needed, we close the valves here or stop the totalizer. This step allows us to put the necessary clean-up routines in one place.

Notice that this sequence does not take the unit off-line, as that coordination will need to be done by either the Start-Next or the Regen state. Doing it this way allows one unit to remain on-line until another unit is on-line; this ensures that there is no interruption of the process flow.

If this were the theoretical world, what we have done so far would be sufficient to run this column. Unfortunately, when building a procedure for the real world we must make contingency plans for most (if not all) possible process deviations. ISA-88 allows us to plan for these contingencies by programming event handlers. These deviations include equipment failures, sensor malfunctions, operator intervention, and process upsets. The consequences of these deviations on the running system (e.g., what happens when the next column is not ready, what if no columns are ready) must be taken into account. For each state of the process, we need to determine how the state responds to an abnormal condition. In a majority of cases the response may be "do nothing." In other cases, a predetermined sequence of events needs to be done to allow the system to operate. To determine what needs to be done, ask the following questions:

1. What conditions would trigger a hold, stop, abort, or other off-normal condition during this state?

2. How should the system respond?

Most of these answers will depend on the operating philosophy of the plant and the specific unit within the plant. In some cases these units are necessary for the continuous running of the plant. In cases like this, hold, abort, and stop are events that must be dealt with quickly. In other cases these units are secondary, and these events can be addressed as time permits. For our three-column example, we have created a table (Table 8.1) that lists the states and the hold, restart, and abort handler actions. (These responses assume a unit that must be kept online.)

After filling out Table 8.1 we can now go and configure the various events for this sequence. Because we can isolate the responses to individual states, each can be individually tailored to meet the needs of that particular state and that specific unit without having to make compromises or complex logic for each type of unit or interrupt. Also, separating the responses makes them easier to understand and therefore maintain. See Figure 8.7 for an example of normal and exception handling procedures.

We now have all the building blocks to complete the batch; all that is left is to develop the recipes for running. In the case of a switching application, the recipe is very simple: start one reactor and continue to run each reactor, one after another. If the units are nearly identical, then the recipes can be redirected to one common recipe. This can include contamination switch amount, impurity fouling rates, and purge cleaning rates. If the units are not identical, then depending on the complexity of the recipe, the next column to be used can be determined by the recipe or can be a recipe parameter.

ISA-88 allows the user to use good engineering practices to design, develop, and eventually maintain the applications. Because we have been able to devolve the procedure into simple and understandable steps, maintenance and enhancements

Table 8.1.   States and hold, restart, and abort handler actions

State name	Hold	Restart	Abort
Initialization	Pause system	Restart where left off	Stop process
On-line	Start next column Isolate current column	Start Regen state	Start next column or stop process
Start-Next	Unknown	NA	Start next column or stop process
Regen	Pause system Isolate column	Restart Regen state	Stop process
Ready	Pause system	Restart Ready state	Stop process

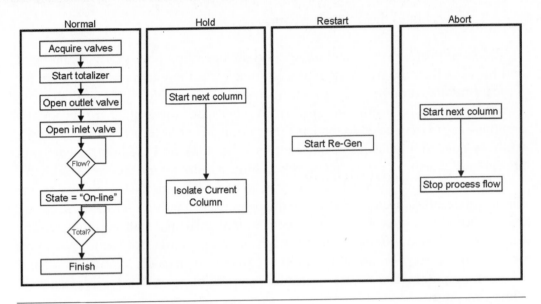

Figure 8.7. On-line state with interrupts.

of these applications can be accomplished relatively easily. When changes are dictated, this structure allows the engineer to isolate the particular state or step or interrupt and then make the changes without affecting the larger application.

Conclusion

While most continuous producers do not like to admit it, they have batch processes. These batch processes typically have requirements to merge continuous with semi-continuous processes. These applications have typically been solved using either complex logic connections or structured text programming languages. These languages are not well suited for interrupt-driven applications. ISA-88 provides the user with a set of good engineering practices. Using ISA-88 based frameworks allows applications to be developed that use interrupt-driven programming. The combination of ISA-88 and interrupt-driven programming allows the developer to generate more appropriate code faster with fewer errors.

Using an example process, we showed how ISA-88 can be used to quickly generate the application. This application generation leads to simple flow diagrams that are easy to interpret, and the resulting code is easier to use and maintain.

Integrating Quality and Process Information in a Batch Context for Semi-continuous Processes

Presented at the WBF
North American Conference,
April 30–May 3, 2007, by

Mike Williams
Water Soluble Polymers Technology Manager
mkwilliams@dow.com
The Dow Chemical Company
1607 Building
Midland, MI 48667, USA

Steve Churchill
Senior System Specialist
sgchurchill@dow.com
The Dow Chemical Company
1607 Building
Midland, MI 48667, USA

Introduction

The ability to meet and verify achievement of customer specifications is a significant requirement for participation in the high-value-added specialty plastics and chemicals businesses. This need is complicated when collecting process information from a semi-continuous manufacturing process. The current ability to monitor, collect, and report conformance to customer specification is impeded by the lack of integrated product quality data with process information in a batch or campaign context.

A process information project was chartered to explore and demonstrate a software application that could provide both process and quality information in a batch context. The product being manufactured is a pharmaceutical excipient, which presents additional requirements for traceability and material genealogy. This chapter will discuss the scope and approach taken to integrate quality information from various analytical devices through a Laboratory Information Management System (LIMS) with a real-time process information system connected to a Distributed Control System (DCS).

Solution Functional Requirements

The following is a list of user and functional requirements for the solution to the LIMS and real-time historian reporting problem:

1. The solution must collect quality information in a batch or campaign context and associate this information with continuously monitored process information from individual units (both batch and semi-continuous) within a combined process cell

2. The solution must be created and supported by the corporate IT function

3. The solution must be viewable globally using a corporate standard desktop solution

4. The solution should automatically collect information from multiple process cells daily and aggregate and store information in a batch context within a centralized and persistent database

5. The solution must enable the statistical measurement and comparison of plant quality performance metrics versus customers' product quality targets

6. The solution should have the ability to filter, drill, and display integrated quality and process information based on any combination of user-defined parameters such as process cell, unit, manufactured date, and product definition

7. The solution must not require extensive knowledge in computer systems applications

8. The solution should provide the capability of generating standard reports

9. The solution should support an export utility to third-party statistical analysis tools

Solution Background

Manufacturing Execution Systems (MES) is a suite of applications that reside architecturally above the real-time process control system layer. The value proposition for this suite of applications is predicated on the availability of real-time information that supports manufacturing decisions. In this particular application, this decision determines the classification of product manufactured, the disposition of non-conforming product, and the corrective actions taken to restore and sustain production of high-quality product for sale.

The objective of this MES application is to unify all relevant process and quality information in a batch context with minimal or no human intervention. By presenting this information in real time, operations personnel have the knowledge to correct deviations from specification, thus minimizing costly downstream reprocessing or disposal of off-spec material.

In addition, batch contextual information by product manufactured facilitates regulatory and customer compliance by generating objective evidence in the form of electronic reports that confirm the integrity of the finished product. This capability is vitally important when serving food and pharmaceutical customers who command a premium price for goods sold. The business expects that this MES solution will be directly leveraged and supported across many similar plants, thus reducing the long-term cost of owning the application.

Some key application features that are critical to business success are as follows:

1. Raw-material use tracking

2. Finished product traceability

3. Electronic batch records and reporting

4. LIMS integration with continuous process information in a batch context

5. Statistical process control of quality-critical process parameters

6. Current batch versus best batch for quality comparisons

7. Automated workflow for final product approval process

8. Validation of process information system design

Solution Architecture

Figure 9.1 illustrates the high-level data flow and product integration for the process information system solution that meets the functional requirements. Process data flows from right to left.

Solution Detailed Design

In this case study, the target facility is composed of five major unit operational areas of which two operate in a batch mode and three operate in a continuous campaign mode (hence the term semi-continuous operation). To place all production in a batch context, a batch data model was created for each area. For those areas that operate continuously, event and time triggers were identified to classify material as it is processed through that unit. By collecting the "batch events," the material can be tracked and eventually traced from the beginning to the end of the manufacturing process.

At the beginning of the process all raw materials are consumed in a batch-converting operation. Inputs are received via spreadsheet entry for manual material loads or through automated sensors for material transfer from bulk storage. After each batch of raw material is reacted, the material is transferred to intermediate storage for downstream finishing through several continuous unit operations. Transfers from intermediate storage to the finishing unit are treated as time-determinant batch events. During the start and end of each batch event or period critical to quality, operating parameters are also collected from the process via a continuous historian. All processing information is combined in a batch context and follows the product as it proceeds downstream to other units and ultimately to final product storage.

At this point, finished product is sampled and analyzed for quality. After 4 hours the quality analysis is posted via a laboratory information management system and added to the processing information associated with the batch identifier. Operations leadership then makes an assessment about the disposition of the finished product. The MES solution facilitates the decision process by querying the database for all information regarding the selected batch identifier and generates a

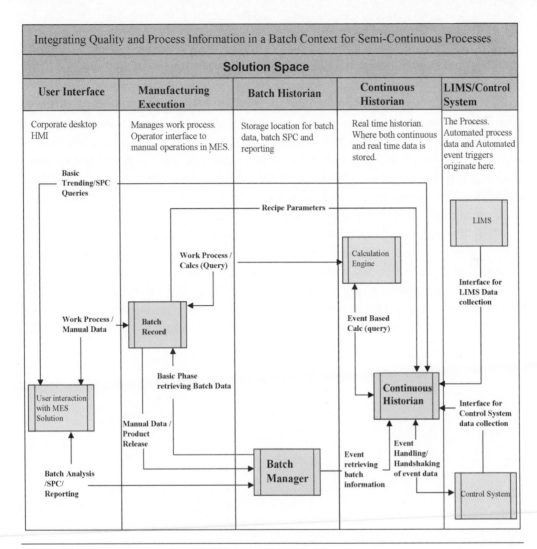

Integrating Quality and Process Information in a Batch Context for Semi-Continuous Processes

Solution Space				
User Interface	**Manufacturing Execution**	**Batch Historian**	**Continuous Historian**	**LIMS/Control System**
Corporate desktop HMI	Manages work process. Operator interface to manual operations in MES.	Storage location for batch data, batch SPC and reporting	Real time historian. Where both continuous and real time data is stored.	The Process. Automated process data and Automated event triggers originate here.

Basic Trending/SPC Queries

Recipe Parameters

Work Process / Calcs (Query)

Calculation Engine

LIMS

Interface for LIMS Data collection

Work Process / Manual Data

Batch Record

Event Based Calc (query)

Basic Phase retrieving Batch Data

Continuous Historian

Interface for Control System data collection

User interaction with MES Solution

Manual Data / Product Release

Batch Manager

Event retrieving batch information

Event Handling/ Handshaking of event data

Batch Analysis /SPC/ Reporting

Control System

Figure 9.1. Integrating quality and process information.

report of compliance to quality-critical attributes. If there are no parameters out of compliance, then the material is released to packaging and sale with an associated certificate of compliance to customer specification. If the material does not pass, then an option is provided for reclassification of the product under a different specification or disposal as off-spec, depending on the quality critical parameters for that batch. Reclassification could entail the blending of various intermediate batches to produce a new batch. Even after blending and reanalysis, the genealogy of the batch is retained to assure compliance with quality specifications.

Benefits of This Solution

The primary benefit derived from this solution is the timely availability and ease of access to batch contextual information that facilitates rapid data-based decisions. In this case, prior to implementation of the MES application, five different people were involved in manually retrieving and evaluating process and quality information from four computer systems and several written logs. Once installed, the electronic batch record became the single source of information, readily available to all decision makers granted security access. The electronic record not only expedites the collection of batch contextual information but also facilitates customer and regulatory compliance audits.

The key business benefit is the ability to direct production in process to expedite customer orders. Another benefit is the ability to recover valuable work in progress that would be wasted due to insufficient records of quality or process condition. The value of this real-time information enables operational excellence, reduced working capital, and on-time customer delivery. It also drives continuous process improvement to meet or exceed customer expectations.

Definition and Format of Recipes for the Packaging of Consumer Packaged Goods

Presented at the WBF
North American Conference,
April 13–16, 2003, by

Andrew McDonald

Global Automation and Control Manager
andrew.mcdonald@unilever.com
Unilever HPC-NA
75 Merritt Boulevard
Trumbull, CT 06611, USA

Abstract

The ISA-88 standard has delivered significant cost savings and benefits to the batch industry worldwide. This has been achieved through standard terminology and models, as well as inherently reusable recipes and equipment phases. The PackML subteam of the Open Modular Architecture Controls (OMAC) Packaging Working Group is a diverse group of end users, technology suppliers, and packaging machine Original Equipment Manufacturers (OEMs) committed to the development of industry guidelines that will deliver Plug-and-Pack® functionality. Their objective is to significantly simplify the integration of packing machinery from different OEMs that are built using control system hardware and software from various technology providers. PackML's work has used the state model within ISA-88.01 as one of its foundations. It has further been discovered that there is significant potential to enhance and transfer the models, terminology, and principles used in batch manufacturing into the discrete world of packaging. The process, physical, and procedural models that appear in ISA-88.01 have direct analogues in the packaging domain. It is also possible to design, in general terms, a recipe that defines both batch and discrete components of a consumer product. The principal advantage to be gained from this approach is the ability to more rapidly deploy and roll out product definition, in a consistent format, to many global manufacturing facilities. This is essential for those global Consumer Packaged Goods (CPG) companies who are committed to ensuring the consistency and quality of their products regardless of where they are manufactured.

Introduction

The objective of this chapter is to clearly show how the principles embodied within ISA-88 can be applied to the discrete world of packaging and why this is important to CPG manufacturers.

CPG Industry Business Drivers

Major CPG manufacturers are committed to growing their businesses, and many have ambitious plans to do so. Revenue growth will be delivered via a continual program of innovation, putting more resources behind a smaller, focused portfolio of brands and pioneering new channels to market. This could mean distribution of products through service providers or creating and running service-based businesses. Operating-margin improvement will provide the fuel for such growth and can be achieved through business process simplification and a highly responsive supply chain.

The consequences of the strategic thrusts of the overall enterprise manifest themselves through a variety of manufacturing pressures (Fig. 10.1). Because revenue growth is a function of innovation, getting new products to market in the shortest possible time is of prime importance. The increased rate of new product introduction means that the life cycle of products will need to be managed more closely. As new products are introduced, older ones need to be discontinued in order to ensure a manageable stable of variations. While companies endeavor to reduce the brand portfolio, the actual variety of products offered increases as brands are extended in different product types. There is continuous pressure to maintain and improve product quality and reduce supply chain costs as well as assets that are used to manufacture them. Clearly any manufacturing operation

Figure 10.1. Packaging pressures.

must be able to satisfy the demand that is created for its products. These pressures require organizations to manage increasing complexity and respond to continual changes, while under pressure to reduce manufacturing costs.

Leveraging ISA-88 in the Packaging Domain

Typical CPG manufacturing sites will have both process and packaging areas. Leading manufacturers have already applied ISA-88 concepts within their process areas. This has provided a high degree of flexibility and has helped manufacturers develop the ability to rapidly introduce and manufacture a wider variety of products, maintain and improve quality, reduce costs and equipment through better asset utilization, and be able to keep ahead of demand. This means that new products and formulations arising from innovation projects can generally be easily accommodated in the process area with few, if any, plant modifications. This is not often true in the packaging area, since packaging lines are constructed to manipulate a relatively small number of pack sizes, shapes, and arrangements. Changing over from the manufacturing of one product to another generally requires removal and replacement of change parts, and this alone is the single biggest loss of time in the packaging of CPG products. Recent developments in digital servo technology have provided greater scope for the design and manufacturing of more flexible packaging machinery. A recent survey conducted by *Packaging World Magazine*[1] indicated that 80% of the packaging machine builders surveyed were applying digital servo technology in their products, but only 14% had specifically redesigned their machines to fully utilize this technology. Such "third generation" machines are being built in order to extract the full benefit that the technology promises. One example of this would be to incorporate the capability for machine functions to be adjusted automatically in response to parameter settings via the machine's control system. These parameters could be encapsulated within a packaging recipe and be used to initialize a packing line. The parameters would be used to adjust the position of servo drives, which in turn change the functionality of the relevant part of the machine. The packaging recipe could also be a component part of the recipe, which defines the manufacturing of the complete product. Thus machine functionality can be changed by setting parameters from a recipe rather than by the manual adjustment of guides and replacement of change parts. To date, there are no published standards specifically relating to packaging machinery control, and little work has been done to define packaging recipes and systems that have the capability to deliver the flexibility demanded by the market. However, many of the ideas, models, and terminology defined within ISA-88 have analogues in the packaging domain. Thus there is an opportunity to leverage this work.

Packaging Machinery Challenges

The packaging machine manufacturing industry is applying increasingly more sophisticated technology to its products and, as a result, delivers greater flexibility and improved performance to its customers. The industry is comparatively fragmented, with a large number of relatively small companies specializing in particular packaging applications. Many CPG companies buy best-of-breed machines from a variety of suppliers and are prepared to accept the need to engineer the necessary interfaces required to integrate individual machines into complete packaging lines. In some cases, CPG companies do not create and enforce global internal standards and specifications for packaging machinery. Their preference is to devolve procurement decisions to local engineering personnel. This means that solutions are very locally focused and difficult to replicate. Moving machines from one facility to another results in significant engineering efforts to integrate a machine into its new location. Some CPG companies have tried to develop their own standards and have required the packaging machine OEMs to comply. This causes the OEMs to build special versions of their standard models (including control system conversion) and leads to increased cost and lead time, lower reliability, and increased servicing costs. Companies that operate in a decentralized manner also experience difficulty getting OEMs to comply with standards because they are not able to leverage their total buying scale.

OMAC Packaging Workgroup

The OMAC User Group is an end-user body originally started by the U.S. auto industry in the early 1990s.[2] The OMAC Packaging Working Group (OPW) is a group of end users, OEMs, technology providers, trade associations, educators, and publishers whose mission is to enhance the value of packaging machinery and systems by promoting the use of digital motion control and OMAC guidelines for open control architectures.

OPW's vision is that over a 3-year horizon, with no increase in capital or operating cost, there should be a 50% reduction in each of the following: delivery lead time, startup time, machine and system footprint, product changeover time, material loss, machine reconfiguration and overhaul time, and machine downtime. In addition, there should be 50% improvements in product throughput, mean time between failure, and flexibility. OPW is organized into five PackTeams, all of which contribute to the delivery and publication of guidelines for packaging systems.[3] The following is a list of the individual teams, with a brief summary of their objectives:

- *PackLearn*. Define the educational and training needs of industry and provide and support programs to meet these needs. This team also works to promote general awareness of this initiative and the industry's move toward digital motion control using OMAC guidelines for open architecture.

- *PackAdvantage*. Identify and communicate with the packaging industry about the benefits and results of using servo motion technology for packaging automation systems.

- *PackConnect*. Define the control architecture platforms and connectivity requirements for packaging automation systems.

- *PackSoft*. Develop programming language guidelines for packaging machinery that will ease learning, support transportability of software across control platforms, and allow continuing innovation by all parties. The vision is to create a common programming language for packaging machinery based on internationally accepted open standards.

- *PackML.* Develop naming convention guidelines for communications between production machinery within the packaging industry.

There is a direct correlation between the business drivers being exerted on the CPG industry and the vision of the OPW. This is illustrated in Figure 10.2. For example, the requirement to reduce changeover time by 50% every 3 years will help reduce time to market for new products, since changeovers will be easier and quicker. Similarly, costs will be reduced because downtime is reduced.

PackML

The PackML subteam was formed in February 2001 and is a diverse group of end users, technology suppliers, and packaging machine OEMs committed to the development of industry guidelines that will deliver Plug-and-Pack® functionality. Their objective is to significantly simplify the integration of packaging machinery (from different OEMs built using control system hardware and software from various

Business Drivers vs. Packaging Requirements

Packaging Requirements	Business Drivers						
	Time to market	Short product lifestyle	Increased product variety	Rapidly satisfy demand	Maintain quality	Reduce costs	Reduce assets
Reduce by 50%							
Delivery time	•	•	•	•		•	
Start-up time	•	•	•	•		•	•
Changeover time	•	•	•	•		•	•
Material loss	•	•		•	•	•	•
Overhaul time	•					•	•
Downtime	•			•		•	•
Increase by 50%							
Throughput		•	•	•		•	•
MTBF		•	•			•	
Flexibility	•	•	•	•		•	•

Figure 10.2. Business drivers versus packaging requirements.

technology providers) so that machinery is delivered, installed, and started in the shortest possible time—the ultimate goal being the achievement of extremely rapid startup of production. PackML's early work used the state model within ISA-88 as one of its foundations. It has further been discovered that there is significant potential to enhance and transfer the models, terminology, and principles used in batch manufacturing into the discrete world of packaging. The process, physical, and procedural models that appear in ISA-88.01 have direct counterparts in the packaging domain. It is also possible to design, in general terms, a recipe that defines both the batch and discrete components of a consumer product.

PackML Modes

The PackML subteam has proposed two classes of mode: User Selectable and Machine.[4]

User Selectable

Two types of modes exist within the User Selectable class: Procedural (in which there are three possible modes of operation: Automatic, Semi-Automatic, and Manual) and Equipment Entity (in which there are two possible modes of operation: Automatic and Manual).

Machine

Machine modes provide the control framework in which the User Selectable modes become available for selection. These can be transparent to the operator. The two proposed modes are Estop and Idle (Fig. 10.3).

There has been much debate on whether the machine modes are, in fact, really states. Since they are invisible to the machine operator in the majority of cases, it has been agreed to consider them as modes. ISA-88 provides guidance on different modes and where these may be used. This work has been leveraged and adapted for use in the packaging machine control domain. The proposed operating modes have been defined along with a relationship model that illustrates transitions between modes. Typical operations, which can be carried out under the procedural operating mode manual, have also been outlined. It is possible that for each manufacturer and each machine, there will be specific state models relevant to operations carried out within the manual procedural modes. The responsibility for the definition of operation will rest primarily with the machine manufacturer. Although some examples of manual operations include Home and Synchronize,

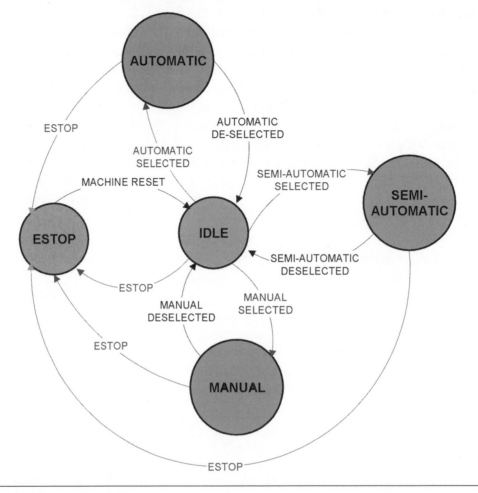

Figure 10.3.   Mode model.

the PackML subteam has already defined a guideline state model for the procedural Automatic mode. Semi-automatic mode will be identical to Automatic, except that the operator will be asked to confirm each transition.

PackML States

The PackML subteam used the ISA-88.01 state model as its basis for design. During a number of reviews, some minor modifications were made to line up with packaging machine control requirements. The final model uses the same state types as ISA-88 and is very similar to the original (Fig. 10.4). A state completely defines the current condition of a machine. A final state represents a safe state (i.e., no moving

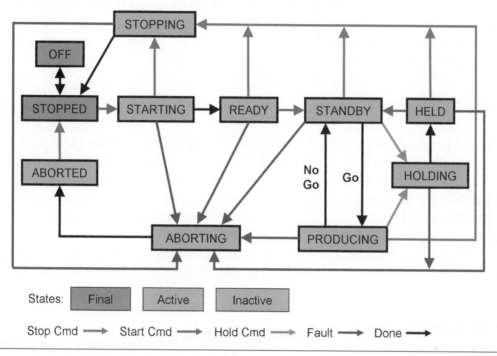

Figure 10.4. State model for Automatic mode.

parts). A transient state is one that represents some processing activity. A quiescent state is used to identify that a machine has achieved a defined set of conditions.

The state model[5] presented in Figure 10.4 has been proposed for Automatic mode; details of the states and the conditions that force transition between states have been provided. Work done to date concentrates on Automatic mode, its state model, and the commands and machine status changes that force transition between modes. PackML has developed mapping functions between its state model and the one defined in ISA-88.01. Packaging machines are generally designed as state machines; the value of the PackML state model is that it offers a consistent set of terminology to be used across different industry sectors. The model is designed to be flexible enough to meet many industry needs and provide a consistent application guideline.

PackML Tags

The PackML tags define a namespace or language for packaging machines. This namespace defines the tag, the data type, and the range of information that is exchanged between machines in the line. This information relates to the current mode and state of the machine and also provides the raw data from which production metrics can be calculated. The first version of the tags definition document has been completed and is available for public download.[6] A selection of tags appears in Table 10.1.

Process Model

For the purposes of packaging, a batch can be thought of in terms of the number of units of production (e.g., a batch of three hundred pallets). For a liquid filling application, typical process stages could be Fill, Cap, Label, Inspect, Carton, and Palletize. For an individual filling machine there could be a number of process operations, such as Infeed, Prepare, Dose, Outfeed, and Quality, each of which is composed of a number of process actions (Fig. 10.5).

These process actions are concerned with controlling specific machine motions, and it is possible to create a number of general-purpose process action definitions that could form the basis of a library of actual software modules. PLCopen[7] is an organization that is actively promoting standardization within the industrial control area. This organization has led the development of a defined library of motion function blocks.[8] Their work has been supported and further promoted by the PackSoft subteam within OMAC. This team believes that the IEC 61131-3 standard is a valuable foundation for structured programming and also believes that the IEC 61131-3 languages support the objectives stated in the OMAC PackSoft Team's mission. PackSoft recommends that technology providers and OEMs adopt the IEC 61131-3 languages for logic and motion.

There is a one-to-one mapping of these function blocks to the library of process actions required for the definition of the majority of packaging applications. Thus it should be possible to define any packaging application in terms of a standard set of process actions, all of which have been defined as standard motion function blocks by PLCopen and adopted by PackSoft. Figure 10.6 identifies the PLCopen motion function block names.

Table 10.1. A selection of tags					
Prefix	Tag name	Tag descriptor	Data type	Units	Range
PML_	Cur_Mode	Current_Mode	Integer		
PML_	Mode_Time	Mode_Time	Time		HH:MM:SS.SS
PML_	Cum_Time_ Modes	Cumulative_Time_ In_All_Modes	Structure		

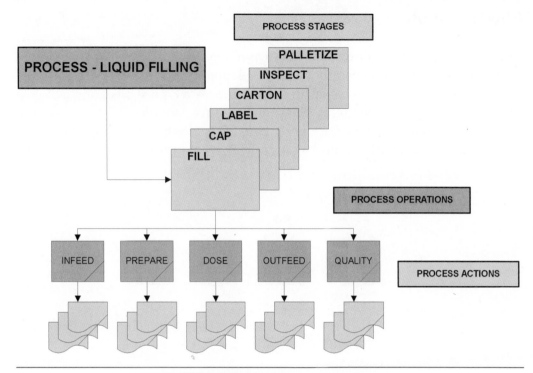

Figure 10.5.   Process model for liquid filling.

Single Axis	Multiple Axis
MoveAbsolute	CamIn
MoveRelative	CamOut
MoveAdditive	GearIn
MoveSuperimposed	GearOut
MoveVelocity	Phasing
Home	
Stop	
PositionProfile	
VelocityProfile	
AcceleratonProfile	

Figure 10.6.   Motion function blocks.

Physical Model

Figure 10.7 illustrates an example physical model for a CPG enterprise. Within a typical site there will generally be areas relating to processing and packaging. Process cells within the processing area will comprise a typical breakdown of ISA-88.01 units, Equipment Modules (EMs) and Control Modules (CMs), as seen in many process plants. For the Packaging area, each specific packaging line can be thought of as a unit within a packaging cell. This conforms with the ISA-88 rule

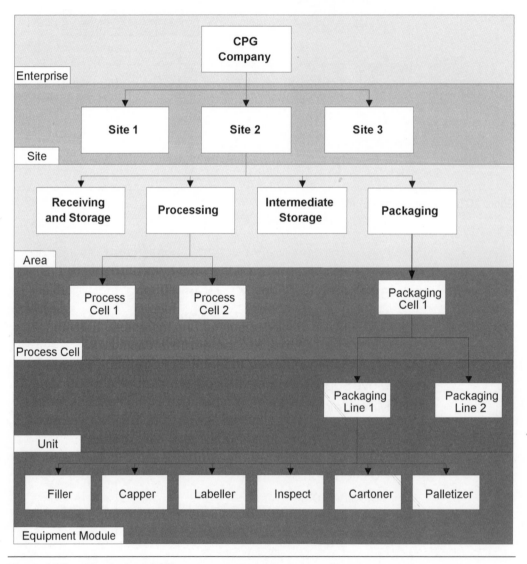

Figure 10.7.   Physical model for a consumer packaged goods company.

that a unit only processes one batch at a time and a packaging batch is analogous to an order. The individual machines within a packaging unit should be thought of as EMs, which are in turn composed of other EMs and CMs. Defining a packaging line as one unit provides additional manufacturing flexibility. The batch produced in the processing area can be stored in the intermediate storage area and then packed on multiple lines simultaneously, since they are units to be acquired by the manufacturing recipe.

Procedural Model

Figure 10.8 illustrates a hierarchical overview for a Distribution Unit or pallet of product. The pallet will be composed of a number of Trade Units (cartons) arranged in a certain manner defined within a palletization recipe. It might also include instructions for the control of the overwrapping operation. There will be specifications for the pallet, the overwrapping material, and labels. There will also be a specification for the artwork that is printed on the labels. This will include the size and position of barcodes and the numbering standard that is used. The Trade Unit is composed of a number of Consumer Units (e.g., bottles of hair care product), a recipe that defines the manner in which the bottles are arranged in the carton, and specifications for packaging materials and artwork. In the future, as on-line case decoration becomes more commonplace, there will also be a need for the definition and execution of a printing recipe.

The Consumer Units (following the hair care bottle example) are assembled from the hair care formulation (according to packaging recipes for fill, cap, label, and inspection); the bottles, caps, and labels defined in the packaging material specifications; and the specification for the artwork that appears on the labels. In the future there could also be the need to execute recipes that control the on-line decoration of consumer units. The formulation is a recipe in the true batch sense of the definition; this defines the manufacturing of the hair care product within the bottle, along with the specification for the raw materials to be used.

The hierarchy in Figure 10.8 shows that the majority of recipes required to make a batch of Distribution Units relates to the packaging and materials handling elements, and it provides an ISA-88 general recipe-type view. This hierarchy will need to be decomposed into site and master recipes for the relevant process and packaging cells. Currently packaging machinery is designed to fulfill specific functions, but the application of servo technology and the guidelines for packaging machinery published by OMAC are being targeted at the delivery of faster, more flexible, and cheaper machines. A number of technology providers (e.g., Elau, Siemens, and Rockwell) have already developed standard modules of control

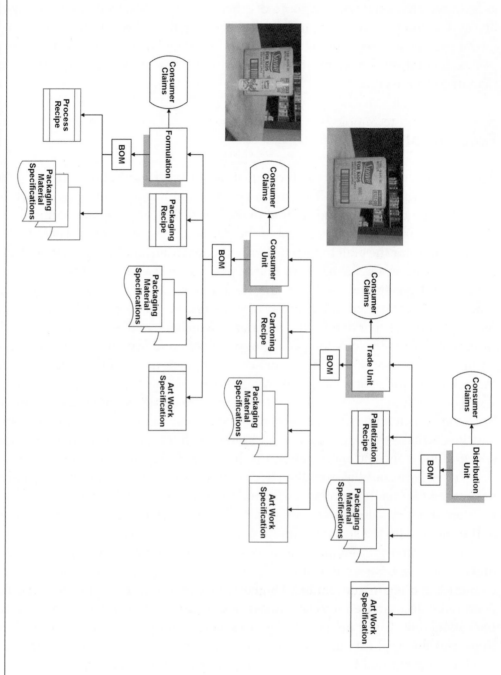

Figure 10.8. Typical product hierarchy.

software that implement the functionality of the PackML state model. The combination of the OMAC guidelines, the motion function blocks defined by PLCopen, and a more modular approach to machine design will result in the delivery of recipe-defined packaging functionality.

Conclusion

The models and terminology defined for batch manufacturing can be applied to the area of packaging. The packaging process can be defined in terms of a standardized library of process actions. This library can be developed through the leverage of the PLCopen standard set of motion function blocks.

A packaging line is analogous to a unit; it is composed of EMs (packaging machines), which are in turn a collection of EMs. This approach fits with the ISA-88.01 definition of an EM (i.e., groups of equipment that can undertake a "finite number of specific minor processing activities"). Such modules are composed of motion control components such as servo drives and encoders. Equipment phases relating to the packaging machine EMs can be constructed using the PLCopen motion function blocks, and a supervisory phase for each machine can be developed in line with the PackML state model guideline. Some EMs (e.g., palletizers) could be shared by more than one line. These can be defined as shared equipment.

A batch of product can be thought of as a number of Distribution Units, as detailed in the product hierarchy (Fig. 10.8). This clearly shows the different types of recipes and how they relate to material specifications. The construction of packaging machines from tried-and-trusted EMs and equipment phases, with flexible functionality defined by a recipe, will deliver significant benefits. For the machine builders and end users, there will be a high degree of simplification.

The engineering required to design and build a packaging machine will be reduced through deployment of standard modules and functional definition via recipes. Constructing flexible machines from prebuilt standard hardware and software components will allow delivery lead times to be reduced. Machine builders will be able to move from the current practice of building each machine to a specific order to an environment that allows machine orders to be fulfilled by the assembly of prebuilt components. Some OEMs might be able to more radically change their business model to one of building to forecast. These developments will enable them to cut delivery lead times, reduce engineering costs and work-in-progress inventory, and increase their responsiveness to customers.

End users will be able to utilize the enhanced machine flexibility to bring products to market more quickly and make better use of their manufacturing assets. They will also be better able to define products in a more generalized format, allowing them to manufacture in different locations across the world. End users in regulated industries will be able to reduce their validation costs through the reuse of previously validated hardware and software components, as well as recipes. This will also lessen the burden on end users' suppliers and will ensure that project costs are more accurately estimated and predictable. The providers of the automation and control products that make all these things possible will see benefits through the ease of integration and more widespread deployment of their products.

This chapter represents an attempt to leverage the great work that resulted in ISA-88. It may not be a completely rigorous analysis, but it lays some foundations for further work.

Recommendations for Standards Committees

Some recommendations for standards committees who would like to apply ISA-88 methodologies to their process are as follows:

- Continue the current proof-of-concept work being undertaken by PackML and extend it to the development of a fully recipe-defined packaging machine

- Investigate the impact of the ANSI/ISA-88.00.03 general recipe standard

- Review opportunities to leverage ANSI/ISA-88.00.02 data exchange and XML schema for information exchange

- Develop a blueprint for the integration of packing line data into enterprise systems through the application of ISA-95 models and terminology

These items can be achieved through closer working between the WBF and the OMAC Packaging Workgroup.

Acknowledgments

Figures 10.1, 10.2, and 10.4 were all produced by *Packaging World Magazine* from figures taken from original PackML documents.

Thanks to the following people for their support, comments, and constructive criticism: Dr. Chris Proudfoot, Unilever; Dr. Fred Putnam Markem, Siegfried Oblasser, Siemens.

References

1. Newcorn, David. 2002. Machine builders speak out on servos. *Packaging World Magazine*, October 2002, http://www.packworld.com/cds_search.html ?rec_id=15074&ppr_key=generation%203&sky_key=generation%203&term =generation%203.
2. OMAC Users' Group. http://www.omac.org.
3. OMAC Packaging Workgroup. Guidelines for packaging machine automation V2.03. January 2002. http://www.omac.org/wgs/GMC/Deliverables/ GuidelinesV2.03.pdf (site now discontinued).
4. OPW PackML Subteam. Machine modes definition document V1.0. February 2003 (site now discontinued, but see http://www.iopp.org/files/public/ WrightSheltonUFlaOMACStandardization.pdf).
5. OMAC. Automatic mode machine states definition V2.1. January 14, 2002. http://www.omac.org/wgs/GMC/SubTeams/PackML/AutoModeMachine StatesV2.1.pdf (site now discontinued).
6. OMAC. Tag naming guidelines. January 2003. http://www.omac.org/wgs/ GMC/MonOmac.pdf (site now discontinued).
7. PLCopen. http://www.plcopen.org.
8. Eelco van der Wal. The PLCopen motion control library: Changing the landscape of industrial control. http://www.plcopen.org/whats_new/plcopen _motioncontrol_LD.pdf (site now discontinued).

Equipment Process Statecharts for Hybrid Manufacturing

Presented at the WBF
European Conference,
November 13–15, 2006, by

Willie Lötz
Manufacturing Systems Consultant
Willie.Lotz@za.SABMiller.com
The South African Breweries
PO Box 782178
Sandton, Gauteng, 2146, South Africa

Abstract

This chapter proposes a generic Equipment Process Statechart (EPS) model with a hierarchical process state extension that may be applied to all levels of the ISA-88 equipment model used in hybrid manufacturing.

The main application of the EPS is to enforce implementation of corporate policy constraints through governance of process and product safety interlocks at the equipment entity level. It supports recipe execution and synchronization of equipment for material transfers to optimize resource utilization and manufacturing efficiency while ensuring uncompromised product quality. Consistent models provide adherence to Good Manufacturing Practices (GMP) at multiple hybrid manufacturing sites and ensures that GMP can be validated and enforced at all times.

Alternative names can be mapped to the standard terminology without changing the basic meaning of the generic meta-model, rules, and relationships. This allows for a more user-friendly presentation in a particular manufacturing domain. Various equipment classes may be explicitly defined by introducing additional substates and rules within the standard composite states.

Using equipment process state models is not a new concept. However, this generic equipment state model provides an abstraction of the business information interface layer to allow reuse of the same semantic model across many different types of hybrid manufacturing process domains.

The model has been extensively used to enhance batch processes in South African Breweries and resulted in significant business benefits and reusability of components. It has now been adapted for packaging-type processes.

Introduction

The globally competitive environment seeks production flexibility to accommodate changes in customer requirements. At the same time, market competition forces more stringent demands of regulatory compliance, validated process safety, and product quality. We have to stringently enforce good corporate governance during all manufacturing processes to address all these growing demands.

We need to deliver more products at uncompromised quality using a minimum amount of resources. Explicit production traceability and increased complexity of the scheduling optimization algorithms applied in sophisticated information systems require knowledge of the actual states of all manufacturing resources.

Modern planning and logistic systems demand accurate and up-to-date real-time equipment utilization information to schedule production orders effectively versus the older "committed, available, and unattainable" plant capabilities. Real-time integration of material quality and process performance information between the production equipment and business systems is expected to support planning optimization and manufacturing performance analyses.

Moreover, the Process Segment Capabilities (defined in ISA-95) of a set of manufacturing resources are not constant for all classes of products, and the resource capability varies over time. We need explicit differentiation in processing capabilities for each class of equipment related to each product class. We must also take into account the effect of the previous usage of the equipment. Scheduling optimization algorithms should use these distinctions in resource capabilities to determine the optimal product mix, sequence of production, and correct material route allocation with the best match to the actual equipment capabilities.

Overall Equipment Effectiveness (OEE) reporting has now become the bare minimum for many manufacturing sectors. Corporate business drivers really expect more effective analysis of actual versus planned manufacturing performance to focus on bottlenecks, optimize scheduling, and narrow the band of control closer to the real equipment capabilities. Sweating the capital assets has taken a much more serious meaning.

A richer equipment state model that exposes the appropriate granularity of manufacturing capabilities at an equipment process states level is required to deliver expanded benefits against more demanding business criteria. A comprehensive range of process capabilities and constraints applicable to each class or individual equipment may be asserted through explicit business rules to declare the expected performance and Quality Of Service (QOS). Process Segment Capability Requirements (ISA-95) should be matched to the resource capabilities by evaluating explicit detail in QOS versus recipe parameters.

Best practices should be described in an explicit way so that they may be reproduced at other corporate sites. Once the best practices have been validated, formal auditing procedures must ensure that the optimized manufacturing performance is sustainable. All proposed changes must be tested and authorized. Actual changes must be immediately identified to ensure the robustness of the total manufacturing configuration management.

Corporate governance of the validated manufacturing environment is now a non-negotiable deliverable in a globally competitive environment. The corporate image and brand integrity cannot afford any slippage. Governance can only be applied effectively across many manufacturing facilities if an agreed upon and unambiguously defined reference model can be used as a compliance type of measurable standard.

Irrespective of the total output, manufacturing facilities use a particular set of units or machines and Equipment Modules (EMs) to produce the product. The large number of different equipment classes that should support awareness of equipment process state in hybrid manufacturing means there is a possibility of huge savings if standardized equipment components could be used. Savings are not limited to the implementation of new manufacturing facilities but extend throughout the life cycle of the facility. Areas of impact include conceptual design, project development, engineering implementation, validation testing, documentation, training, production operation, data gathering, performance reporting and analysis, plant maintenance, and process upgrades. The critical impact of proven components is the ability to design once and implement many times without revalidation.

To maximize our current investment in manufacturing automation and information systems, it is important that new methods of modeling and implementation provide backward compatibility as far as possible. Various levels of automation at different sites should be moved toward a consistent global manufacturing framework, and the changes to these sites should be supported using proven components.

Business expectations have become more sophisticated, and corporate governance is non-negotiable. Proven standards can provide a stable platform to extend the system functionality without reinventing the wheel.

Standard ISA-88.01 Equipment State Transition Model

The well-known ISA-88.01[1] standard defines a state transition model for automated control execution. It has served the batch process community very well for many years, and it is an excellent basis for the new wave of enhanced manufacturing processes (Fig. 11.1).

Control Modules (CMs) that implement this model would directly execute a valid command without considering the business rules of the manufacturing process related to material flow control. An individual valve should open or close when commanded in a manual mode, without caring if the route on either side contains incompatible materials such as beer and caustic soda used for cleaning the route.

Maintaining process integrity is only possible when that valve is configured as a permanent part of an EM that defines a specific volume within the equipment that must be protected from external influences during production. ISA-88.01 control recipes use equipment procedural control to interact with the control devices.

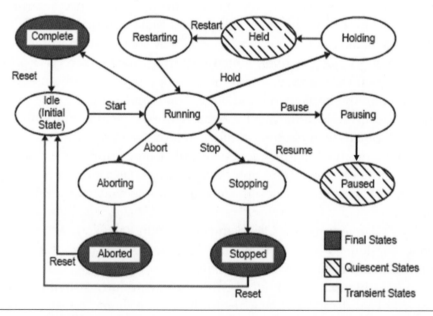

Figure 11.1.   ISA-88.01 equipment state transition model for automatic control mode.

If process validation and protection is built into the normal equipment procedural control, such as equipment phase logic, the exceptional process conditions possible during manual or semi-automatic control mode could not be validated, and process safety or material integrity could not be enforced in all situations.

We need an independent policeman at the level of basic equipment control. It must be built into all controlled equipment to enforce material integrity, with the abilities to execute some local validation of the requested change command and to either allow or disallow a command in all modes of control.

Evaluating embedded business rules of the EM should enforce certain rules such as the concepts defined by "Flow Breaking Control"[2] before allowing execution of the requested command.

Proposed State Model for Packaging Machines

The joint workgroups of the Organization for Machine Automation and Control (OMAC), ISA, and the WBF packaging community are developing a proposed standard state model for packaging machines that is similar to the ISA-88 batch control state model. The current discrete machine model has evolved from the original packaging machine version and is much closer to the original ISA-88.01 batch control state model. The crucial difference in the discrete processing end user requirements is that we are now explicitly dealing with a machine that operates at a significantly higher level of control in the equipment hierarchy than a CM or EM, where the original ISA-88 state model was targeted (Fig. 11.2).

Why Do We Need an EPS Model?

Higher levels of entities in the equipment model must now deliver operational business value and reflect the equipment process state rather than provide only the detail technical state of low-level components. The equipment entity must also respond effectively to higher-level commands or events produced by an external controlling entity such as a control recipe.

The focus of process cells, units or machines, and EMs must be on optimizing manufacturing performance rather than providing technical detail of engineering components that are irrelevant to the perspective of the Operations management level.

As an example, the ISA-88 state model cannot directly reflect the "Dirty" or "Clean" process states of an EM or unit. Transition logic cannot be defined to manage the flow decisions associated with these external events that affect the equipment process state (Fig. 11.3).

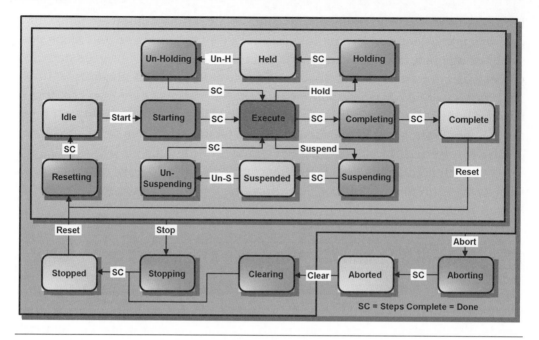

Figure 11.2.　PackML state model version 3.0 for automatic control mode.

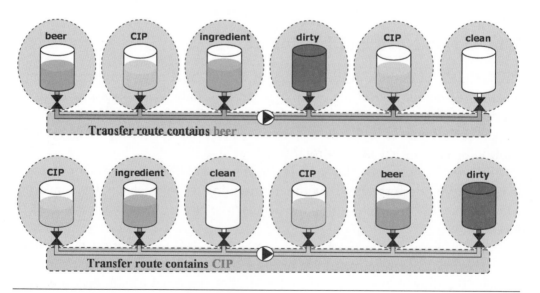

Figure 11.3.　Typical material transfer pitfalls.

Most batch and continuous manufacturing processes are "wet" types, but there are also some packaging machines such as bottle fillers that would be classified as a wet type. After completion of a batch or filter run, the units or machines and wet material transfer routes potentially carry material residue that would cause unacceptable product contamination of the next batch unless corrective action was taken.

Basic equipment state models may be extended to provide wet types of units or machines and EMs with process state information and accompanying business rules to validate flow sequences. Even for dry types of equipment, adding process state information provides major business improvement opportunities such as direct support of OEE. The process state extension can be done without violating the current ISA-88 state model.

Developing an EPS Model

The basic equipment state model may be extended in a hierarchical structure to support all levels of equipment effectively. Composite states would contain the substates and "roll up" the information in each successive layer via embedded business rules.

Instead of the flat structure of the original ISA-88 state model, a hierarchical type of statechart model may be used to focus attention on the relevant issues while not loosing the context of underlying flow structures. To demonstrate backward compatibility, we shall first show the step-by-step development of the statechart concept using the initial ISA-88 model. Then we will use this technique to extend the model.

Reviewing the ISA-88 State Model as a Statechart

The initial basic equipment control state must be established first in the hierarchy. This sets the context and relates the initial state to the startup sequence for the equipment. The associated equipment controller (typically a locally embedded Programmable Logic Controller [PLC]) establishes an instance of BasicControl (Fig. 11.4) when the equipment controller successfully completes the controller PowerUp event.

The equipment BasicControl layer must have an embedded finite state machine with an algorithm to manage and control transitions between all states. Only one state can be active at any time. The PowerUp event forces control to a Terminated state (Fig. 11.4).

Correct functioning of equipment BasicControl is absolutely essential to support all the successive layers of control. Without active BasicControl, neither the

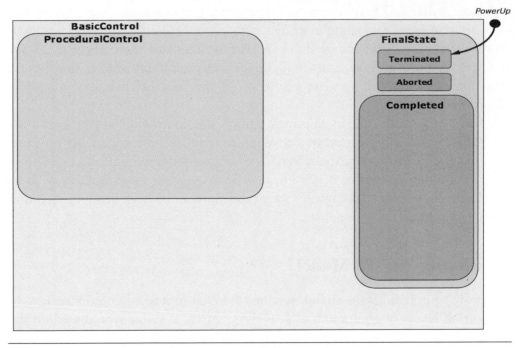

Figure 11.4.   BasicControl enables definition of ProceduralControl.

FinalState nor ProceduralControl (or any of their respective substates) can be established or sustained.

Because CMs are managed at the level of BasicControl, the current state of actuators and embedded regulatory control loops of the CMs will be maintained. The unit, machine, or EM is thus in a stable control mode, waiting for the first external command.

Mechanical and electrical "health" of the equipment entity is continuously monitored by BasicControl. If BasicControl is deactivated for any reason, all the equipment control effectively "dies." Note that the "E-stop" event can only be recognized after BasicControl has been successfully established (Fig. 11.5).

The ISA-88 concepts of BasicControl and ProceduralControl are shown in a hierarchical equipment statechart model (Fig. 11.6). Although the ability to establish ProceduralControl functions has been enabled through the BasicControl layer, Producing has not been initiated and equipment phase control is still inactive.

All other substates in the NotProducing state also only require correct functioning of the equipment BasicControl function to be available. The exit or reentry events and states in NotProducing are common functions that may be invoked from any of the ProceduralControl functions. The events may be parameterized

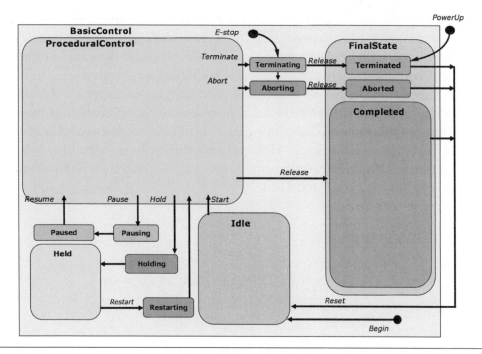

Figure 11.5. BasicControl also enables definition of exception states.

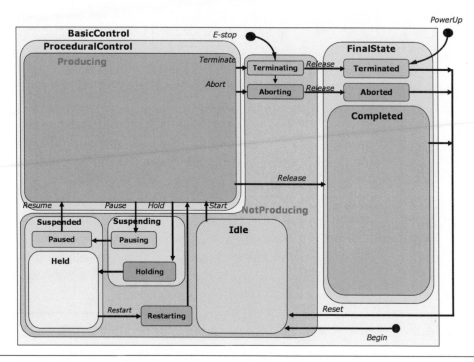

Figure 11.6. ProceduralControl enables Producing.

according to predefined rules per type of state to customize the flow of control for further processing, contextualized program reentry, and historical event recording.

Prior to reaching actual Producing or when production is Suspended, the equipment will be in a NotProducing state. Historical data with the period spent in each state can be used to easily calculate OEE. The only difference to the original ISA-88 state model is that we have added some underlying context of powerful composite states to easily reflect the active contribution of the equipment to the business value of manufacturing. Hiding some bottom layers of context reduces clutter; it is now clearly evident that the basic structure of the ISA-88 state model has been retained (Figs. 11.7 and 11.8).

Generic Flow of Producing

A generic set of high-level composite states that will add significant business value is proposed as the core of the extended equipment process state model. We need a minimum number of states that will support consistent system semantic

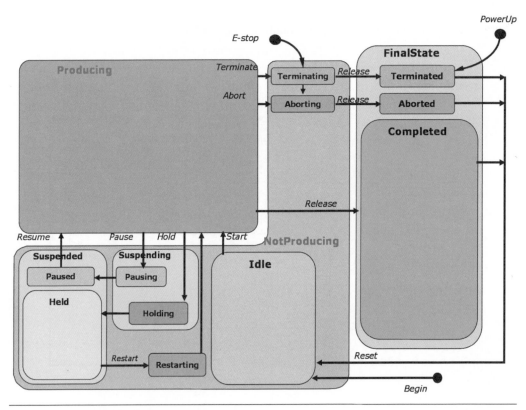

Figure 11.7.   Statechart of generic control.

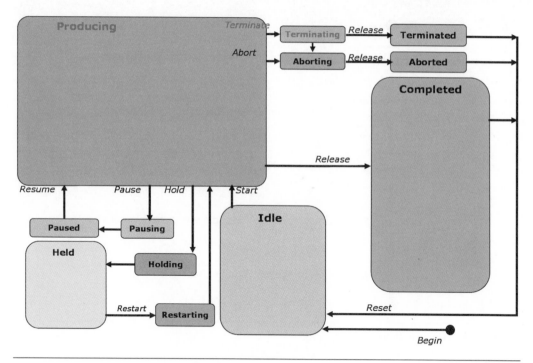

Figure 11.8. ISA-88 equipment state transition model in statechart format.

interoperability, based on a concise equipment process state model. The states of all participating units, machines, and EMs could then be reflected in a uniform framework at the Process Cell level. Equipment status overviews with direct drilldown capabilities based on real-time information are powerful tools to effectively manage and optimize the coordination, planning, and utilization of manufacturing resources.

The Producing flow always follows a basic sequence to define the interaction between the "Actor" (equipment ProceduralControl) and the "Object" (material and components) to manufacture the "WorkProduct." The sequence corresponds to the "major processing activities in a Unit," as described in ISA-88: "An operation . . . defines a major processing sequence that takes the material being processed from one state to another, usually involving a chemical or physical change."[1] That is, it defines the high-level operations of manufacturing from a generic perspective. This generic Producing sequence can be standardized (Fig. 11.9).

Process flow for a unit is defined by the execution of a unit procedure as part of a control recipe, according to ISA-88. When we talk about the sequence of Producing states in the EPS, we refer to the effect on the physical status of the equipment due to the execution of the operations. The operations are thus the cause of the

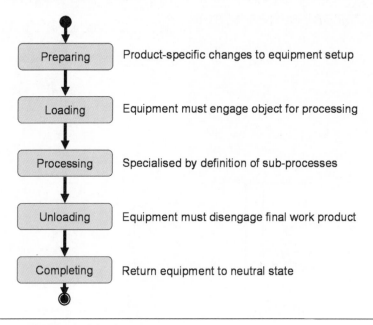

Figure 11.9. Generic flow of Producing.

equipment process state change events. There is a direct relationship between specific operations and equipment process states, but they are not the same things.

We focus on equipment states that frequently result in production delays due to practical external effects such as material starvation or buildup. Predefined business rules can identify internal equipment procedural problems and cause other failure events. When significant periods of inaction can be identified quickly (either in real-time or historical analysis), then corrective action can be initiated to minimize lost production time.

Delays sometimes occur between the equipment states where some required activity takes place and the next valid activity. It helps to define additional equipment states that reflect the "done" state for the critical sequences, where delays occur to extend the "done" state beyond the expected or allowed delay. These states can all be presented visually to the operator, but electronically monitoring these periods of wasted time and generating automatic operator alerts according to business rules helps minimize the loss.

Standardization of the EPS at this level of detail supports structured, efficiently engineered implementations and consistent validation of all generic functions of generic components. Detailed transition rules can be defined by the end user. Some manufacturing processes are simple and do not need great detail, while others may require many levels of detailed substates and complex transition rules.

The flow of batch processing, discrete processing, and continuous control processes involve one or more machines or equipment entities with defined equipment control capability. Some manual interaction may also be required as complementary Actors.

The ability to add substates within each of these composite states provides versatility to define any equipment class to the required level of granularity without invalidating the generic statechart. Specialization of the substates means that only the additional programs need to be tested while "inheriting" the established generic functions.

The use of substates provides equipment model extensibility while still complying with a standard for EPSs that supports generic flow information applicable to hybrid manufacturing plants. Implementation overhead can be minimized by using simple transitions to immediately skip through the state if the process does not require detail of the substate. However, it would be a good practice to build the structure for all the standard states. This enables possible future enhancements to take place painlessly.

Generic Flow of the FinalState

It is advantageous to define generic substates of the Completed state to enable implementation of some standard business rules that are generally used to cater to practical effects on the QOS influencing the equipment capability. This includes physical equipment wear and tear or other types of performance degradation such as natural, chemical, or biological processes affecting hygienic cleanliness. Transition rules may be defined for each class of equipment that will force a ServiceExpired state (Fig 11.10) when operating under certain conditions. Equipment is then Unavailable and locked to prevent normal production until the equipment has been correctly Serviced. Manual intervention or automatic scheduling may set the ServiceRequested state.

There are many possible kinds of Servicing, depending on the circumstances (Fig 11.11). Servicing may be executed as an external manual or semi-automatic procedure. It can include mechanical refurbishing, lubrication, instrument calibration, performance testing, and other kinds of servicing.

Automatic service execution such as Clean In Place (CIP) may also be accomplished via an automatic execution of predefined equipment ProceduralControl logic. This would be a kind of cleaning recipe, similar to a production recipe but using different formulae and rules. A special kind of ServicingReset is then required to select the appropriate transition rules, but the generic substates from Preparing through Completing are still valid for a CIP recipe. The result (product) of successful Producing is clean equipment. After Completing an automated

ServicingRecipe, the equipment returns to the AvailableServiced substate to enable more Producing of actual product.

Whenever equipment is in the Aborted or Terminated state, it means that some exceptional events forced an abnormal process completion. The state of the equipment is really indeterminate, and it should automatically force a ServiceExpired state to ensure proper reinstatement of working conditions. These business rules applied to the FinalState are very handy to enforce strict governance of batch-process-type equipment or any wet process equipment such as packaging filler machines.

Generic Flow of the Held State

A Hold event may be initiated by an external Actor such as an operator or higher level of ProceduralControl. Producing may be resumed if there is no fault by the Actor issuing a Restart command.

When a Hold event is automatically generated during Producing due to some process interlock or other detected problem, the Failure event should carry the context of the fault to assist in diagnostic procedure and corrective actions. Human intervention is required to clear and reset the fault under controlled conditions. Only when the Failure flag has been reset by clearing the root cause will the flow management system allow a manual Restart command.

Extending the ISA-88 State Model as an EPS

We can now use the Producing substates to extend and specialize the original Producing state. This will reflect the influence of the actual manufacturing process on the state of the equipment. We could apply the statechart to all levels of equipment from the EM to unit or machine level and even higher.

During the Producing state of the equipment, the actual manufacturing period while Product is being produced should be distinguished from the PreProduct and PostProduct states. Time lost in NotProducing and FinalState pinpoints the manufacturing issues that don't contribute directly to the bottom line, even though the steps may be essential for specific reasons. Industry benchmarking of similar processes and analysis of each aspect enable optimization of utilization of manufacturing facilities.

As shown, other states may be extended by substates where the detail will add business value. The model shown in Figure 11.10 is probably the lowest that we can go while still maintaining a generic perspective of the statechart applicable to any class of manufacturing equipment.

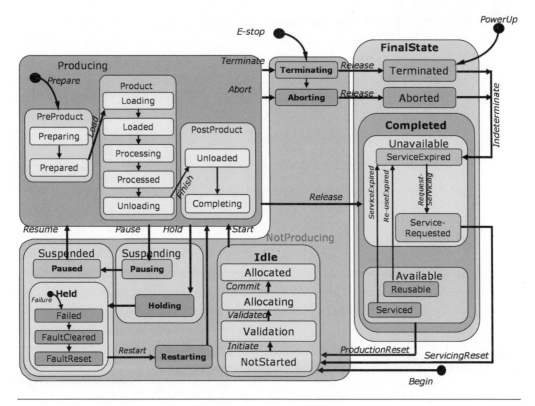

Figure 11.10. Extending the ISA-88 model toward an equipment process statechart.

Examples of Embedded Business Rules

Business rules may be specified per class of machine by defining parameters with downloaded values for engineering configuration. Parameters may also be product related. For instance, some classes of products require special servicing more frequently than other production processes.

All the current accumulated periods and cycles, as well as the allowed target values per parameter, must be stored locally in equipment that is accessible to the flow management system.

The local controller manages the equipment process states according to evaluation of the current state and requested state change event, as specified by the defined business rules. Some examples are listed in the following sections.

ServiceExpired Event

This event, shown in Figure 11.10, may be caused by any of the following:

- Reaching a preset time-out period or number of cycles of normal manufacturing usage

- Machine lubrication

- Replacement of items due to normal wear and tear (seals, bearings, etc.)

- Instrument recalibration or mechanical alignment

- Equipment performance validation testing

- GMP or corporate policies to ensure effective cleaning

ReuseExpired Event

This is similar to the previous case, but may have separately defined rules or parameter values. Business rules are embedded in a local flow management system that will monitor equipment performance, validate the correctness of external commands to request different process setup or state changes, and enforce GMP in general. The Equipment State Manager is the guardian angel of the equipment, process, and materials.

Using Equipment Process State Cycles

Engineers would use the extended EPS to define, develop, configure, test, and diagnose problems at a technical level of detail equipment control. This level of detail is irrelevant to an operator and the cluttering effect can be hidden without being lost.

The Human Machine Interface (HMI) presentation is simplified to focus on the critical Process states during which time-consuming delays may happen or to show exception states that would reduce the overall manufacturing performance. Clear visibility of all critical process states supports the operator in quickly analyzing and rectifying production problems (Fig. 11.11).

A complete manufacturing cycle also shows the normal servicing requirements of the production equipment. The normal flow pattern should follow a clockwise path along the outside ring of states of the model. Whenever any of the states not within the U-shaped Producing composite state become active, manufacturing losses are occurring. Multiple equipment statecharts would show the upstream and downstream states that might delay optimum material flow along the path.

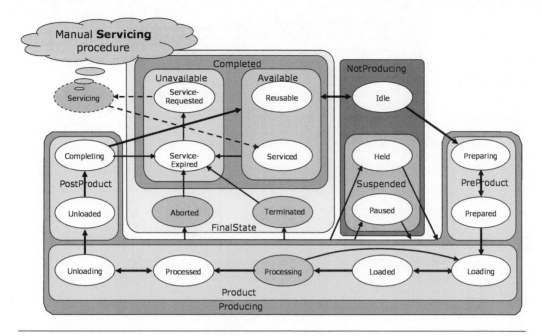

Figure 11.11. Generic equipment manufacturing cycle.

Ineffective scheduling reducing optimum utilization of equipment would be evident when equipment spends too much time in the NonProducing or FinalState in relation to the active Producing period.

Specific Equipment Statechart Models

The state names used in the generic Equipment State Cycle might be confusing in a particular manufacturing community. Appropriate domain names for states and events could easily be mapped to the standard model and terminology as long as the alternatives are semantically equivalent to the standard.

Any or all of the states may be enhanced by using additional substates and appropriate event definitions, as long as the flow structure is not compromised. In most cases, the generic Processing state does not provide sufficient granularity to really follow the critical manufacturing process sequence. Additional substates may be defined within Processing to specialize the process states for a particular equipment class (Fig. 11.12). In a batch process, this additional level of substates would follow the naming for unit operations for all the transient states (i.e., states ending in "ing" where work is being done).

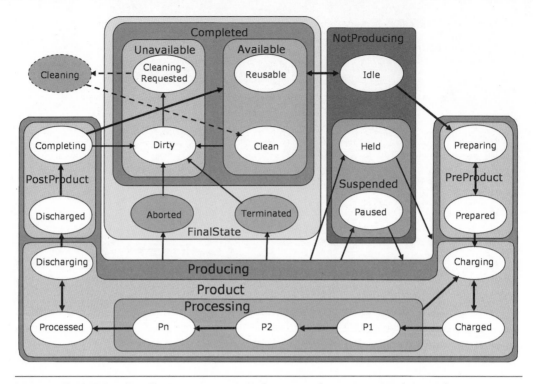

Figure 11.12. Extending the Processing state for a specific equipment class cycle.

Conclusion

We can conclude the following from our experiences:

- We can define a generic equipment state machine applicable to all classes and hierarchical levels of Equipment, including batch, discrete, and continuous control
- The original ISA-88 model is extensible to reflect equipment process states without deviating from the basic principles
- Current existing equipment state models are backwards compatible with the proposed EPS
- Specific models for different equipment classes can be achieved by definition of substates and transition rules
- Alternative state and event domain names (even other languages) may be mapped to the standard namespace, provided they are semantically equivalent

- Embedded equipment state transition conditions enforce business rules to ensure equipment safety, process interlocks, and material integrity in all events
- South African Breweries will work via the ISA and WBF workgroups to further develop and standardize the extended model for a generic EPS

Addenda

Figure 11.13 shows an implementation of the State Manager pop-up for equipment AT001.

References

1. ISA-88.01–1995. Batch control part 1: Models and terminology.
2. WBF-WG. ISA-88 modeling using flow analysis, draft 3.0.

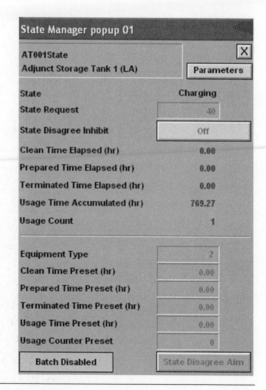

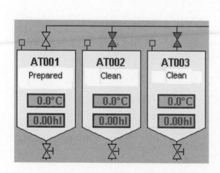

HMI Implementation

Figure 11.13. Example of equipment state manager pop-up.

Process Meets Discrete: Coupling of Batch Control to OMAC and to Real Packaging Lines

Presented at the WBF
European Conference,
November 13–15, 2006, by

Eelco van der Wal
Managing Director
evdwal@plcopen.org
PLCopen
PO Box 2015 NL 5300 CA
Zaltbommel, Netherlands

Abstract

The processing of a product is normally not the final stage; the product needs to be packaged. This is the area where process meets discrete. With pressure on the marketing of the final product, suppliers of consumer goods fight for shelf space in the supermarkets. Many consumer marketing campaigns are done for this purpose, which either add something to the product or give a percentage for free. In all cases, it involves changes in the packaging lines. And these changes need to be made in a flexible and fast way, while still fitting in with the overall concept.

Currently, there are certain activities going on to link the process side of production to the packaging of the final product that is ready for shipment. This presentation shows how this coupling can be done.

Introduction

ISA-88 and ISA-95 are well-known standards in the process industry. However, in many production processes the processed result is not the final product. For this it has to be converted to a packaged product in a wide variety of forms. It is this area where process meets discrete.

The discrete environment, especially in packaging, is supported by the Organization for Machine Automation and Control (OMAC) Packaging Workgroup. This workgroup relies heavily on the IEC 61131-3 standard and the PLCopen additions to this. This chapter will show how these all combine and create an excellent basis to merge both environments effectively.

State Diagrams: A Matter of Vision

Both ISA-88 and the OMAC Packaging Workgroup have defined a state diagram. There are differences between the definitions, but they are well documented and activities are going on to harmonize these two. The OMAC state diagram can very easily be mapped on the IEC 61131 structured language tool called a Sequential Function Chart (SFC). An example on how this is done is available at the PLCopen Web site.

Integrating the Packaging Production Line

A packaging line consists of multiple functional units like primary packaging and final packaging, and it may include palletizing functions. If we want to couple a packaging line to the batch model, we need to make sure that the application software for the different functionalities in the packaging line supports the concept of recipes. This may look straightforward, but it is not; the process side and discrete side need different information. The discrete side needs information on the packages to be used (e.g., the language in which they are printed, the size to be used, the number of packages in a box, and the number and order of the boxes on a pallet), as well as the correct shipment information and the type of expiration date of the packaged product. See Figure 12.1 for a typical packaging line arrangement, where "unscramble" means transforming a random arrangement of bottles or caps into a linear feed to the machine with each bottle or cap in the proper orientation. This of course requires adaptations in the batch recipe model and also requires a structured approach to the application software development.

The worldwide IEC 61131 standard provides the necessary software tools and structured approach to cope with this. Based on this, PLCopen has harmonized the motion control functionality that is essential to all packaging machines.

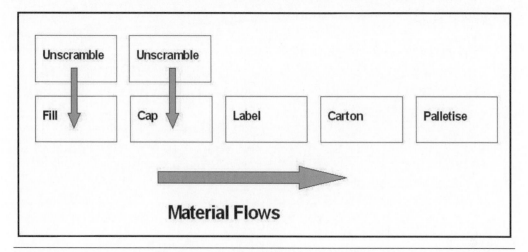

Figure 12.1. Typical packaging line arrangement.

With this PLCopen motion control profile, a more flexible approach is possible in the packaging machine. It provides the basis for a mechatronic approach in the development of the packaging machine, which has gained wide support from the industry. It now supports a variety of control architectures as well as suppliers. With support for simple motion movement, as well as functionalities like camming and gearing, it basically can fulfill most packaging application needs. Figure 12.2 shows how a slaved servo motor may generate a cam position function from a master position signal. This replaces the mechanical master line shaft and the gears and cams required to generate the same motion.

The PLCopen motion control specification uses an axis-related state diagram that shows the status of each axis. There is no contradiction between this lower-level state diagram and upper-level models like those defined by ISA-88 and OMAC. Actually, they combine very nicely. An important addition to the state diagram is related to the error handling of each axis and so on for a full machine.

An example of the implementation of a mechatronic approach using IEC 61131 as well as the PLCopen motion control is shown in Figure 12.3, which illustrates a flow-wrapping machine. For the flow wrapping, three motors are used that are coupled via software to provide flexibility that cannot be realized in a mechanical solution.

Overview of PLCopen Motion Control

With ever-increasing processing power, different motion control solutions are possible: both centralized systems and distributed systems, based on multiple

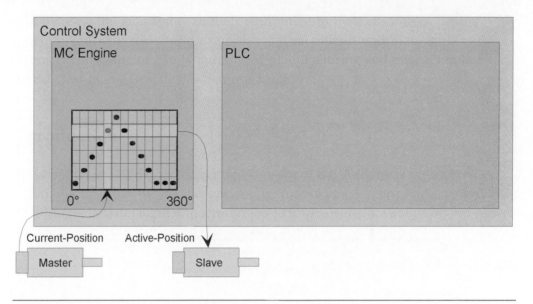

Figure 12.2. Example of the PLCopen camming functionality.

intelligent drives. This changing environment, combined with the "classical" Programmable Logic Controller (PLC) environment integrating more and more motion control into their controls, created a basis for standardization: let us not all invent our own wheel but instead do it together. This vision is currently shared among many different suppliers as part of the PLCopen organization and has resulted in the definition of a PLCopen Motion Control Library. It is a standard in the programming environment to harmonize the access of motion control functionality across platforms. This library has changed over time as shown in Figure 12.4.

Effectively, this standardization is done by defining libraries of reusable components. In this way the programming is less hardware dependent, the reusability of the application software is increased, the cost involved in training and support is reduced, and the application becomes scalable across different levels of control. As such, it is based on IEC 61131-3 function blocks. With the standardization of the interfaces and the functionality and implementation on multiple platforms, the generated application program is much more hardware independent and thus reusable across platforms. Due to the data hiding and encapsulation, it is usable on different architectures, ranging from centralized to distributed control. As such it is open to existing and future technologies. Overall, the standardization is expected to cover around 80% of the motion control market. Currently, the PLCopen Motion Control Specification consists of the following five parts:

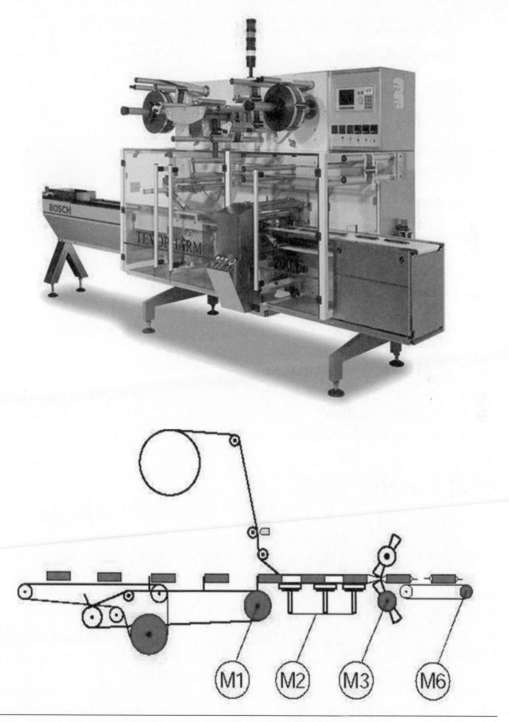

Figure 12.3. Example of a flow wrapper with synchronized multiple axes.

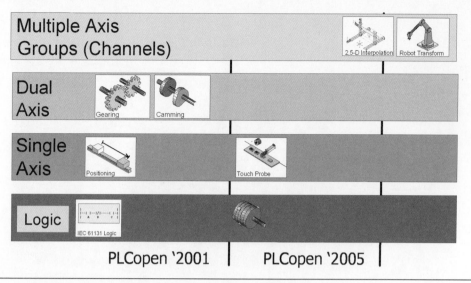

Figure 12.4. Positioning of the PLCopen motion control library over time.

- Part 1: Basics
- Part 2: Extensions
- Part 3: User guidelines
- Part 4: Interpolation
- Part 5: Homing procedures

OMAC PackML

The packaging and converting production lines consist of a set of functionalities (e.g., machines) that are interrelated. This relationship is reflected in the OMAC state model. Through the implementation of this state model on all relevant functional units, the operation of the machine as well as the relevant process data is made available to the user in a standardized and simplified way. This is known as PackML, which is short for "packaging machine language." Simply said, if machines could speak they would speak PackML. With this, a packaging and converting line can be looked at from a higher level (Fig. 12.5).

Mapping of OMAC PackAL

Within the OMAC Packaging Workgroup, a specification has been released that deals with packaging functionality on a higher software level called PackAL. This specification relies heavily on the PLCopen motion control set of specifications.

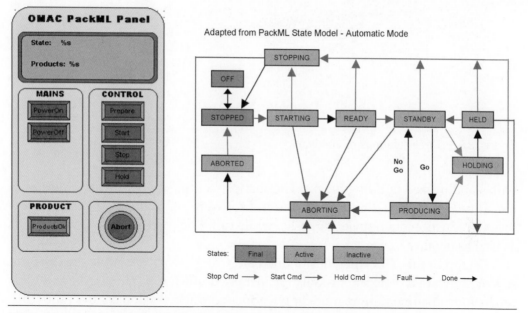

Figure 12.5. Example of an implementation of the OMAC PackML state model.

However, this does not mean that new function blocks have to be defined to implement the specified functionalities.

Within the third part of the set of PLCopen motion control specifications (which deals with the user guidelines), examples are under construction to show how these generic functionalities can be met by the existing defined motion control function blocks. With this, the application programmer can define the block's own functionality within its own company-wide, user-derived function block library. This flexibility provides support for different philosophies in the sense of different control algorithms and sensors. For example, the winding and unwinding functions can be solved with different solutions. OMAC specified one of these, which can easily be mapped to the PLCopen provided functionality. However, more solutions for this basic problem are possible. These solutions can differ in their sensors, actuators, or control specifics. By creating their own library, users can be flexible in their solutions while still providing a similar functionality in the function blocks in their application program (Fig. 12.6). This is beneficial for the operators as well as the maintainers. Figure 12.7 shows an arrangement of standard function blocks and "glue logic" for constant product velocity winding as the roll radius increases.

Other PLCopen Additions

Besides motion control, PLCopen also has created a standard dealing with machine safety. With this the safety functionality is provided to the user on a software level

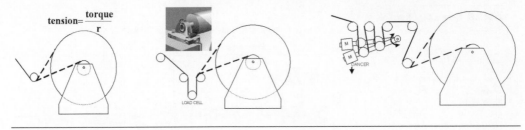

Figure 12.6.　Three solutions to winding.

with standardized building blocks. The changeover for safety functionalities from hardwired systems to digital networks and software is provided. This adds to the overall reduction of engineering tasks, while fulfilling increasing safety demands as well as liability issues.

On top of this, PLCopen has defined an XML schema to couple the different sets of software development tools for increased efficiency. These different tools can include documentation, modeling, verification, and version control. With this, the IEC development application environments are no longer isolated but can be integrated in a full suite of relevant software development and engineering tools. This, again, is a major contribution from PLCopen to the community.

This usage is currently under consideration to play an essential role in defining and generating the application software for complete production lines in the automotive industry with up to 40,000 motors and drives by the major players. Thus, one can be confident that it also can be used within much less elaborate packaging and converting lines.

Conclusion

Currently, there are certain activities going on to link the process side of production to the packaging of the final product, ready for shipment. The OMAC Packaging Workgroup with activities in PackML, PackAL, and Make2Pack provides a framework to handle batch concepts throughout the entire production side—from raw material to packaged goods ready to be shipped.

OMAC makes references to PLCopen (http://www.plcopen.org). This association has merged motion, logic, and safety onto one development platform. Moreover, it provides a harmonized interface with the discrete control side, which can be accessed via PLCopen XML. Using these constructs, the coupling of process and discrete including batch recipes can be realized.

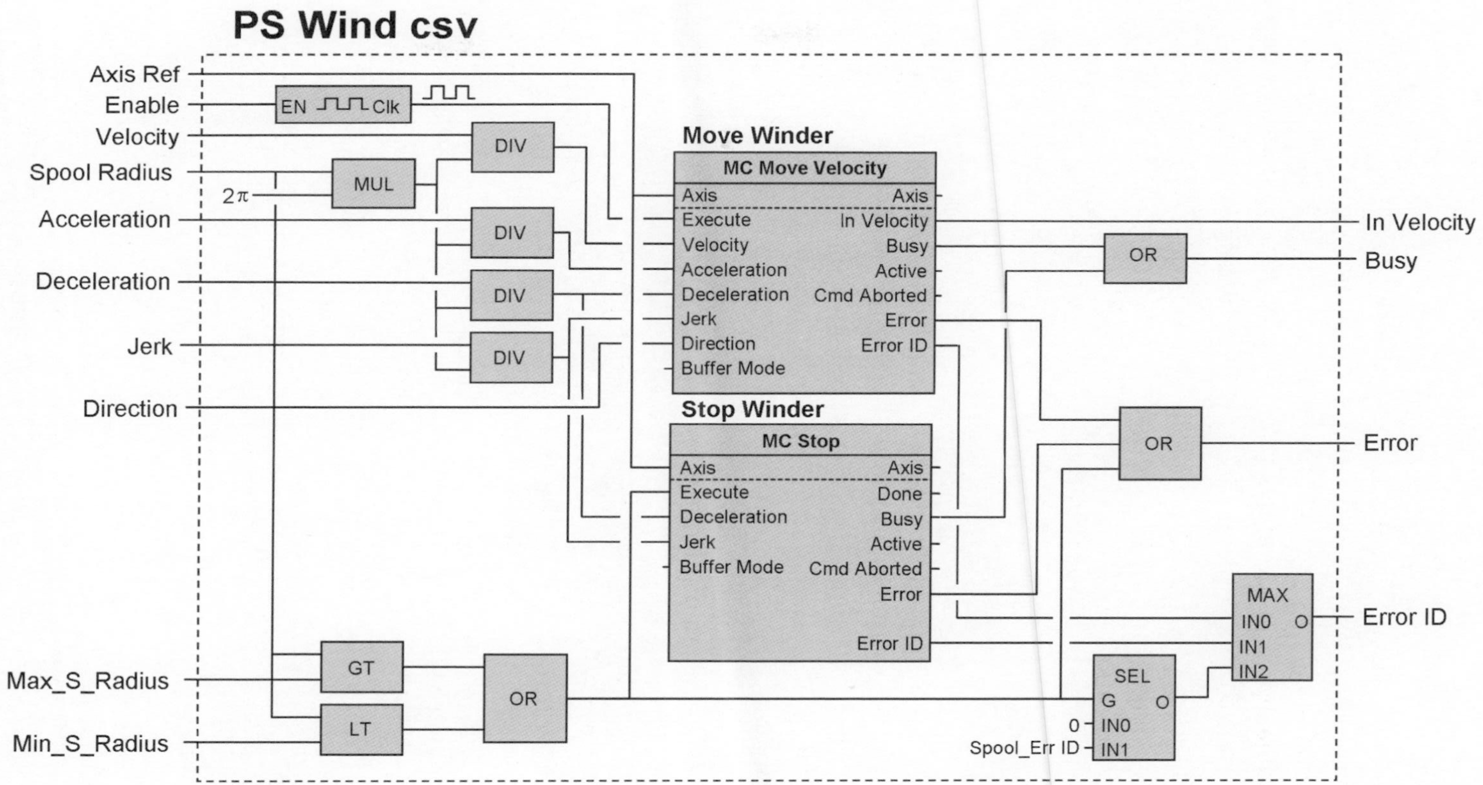

NOTES :

- As the spool radius is changing the units of MoveVelocity are defined in rev/sec.
- Other possibility is injecting the SpoolRadius into the axis scaling parameters, but this is more vendor specific operation.

Figure 12.7. An example of the mapping of the defined OMAC winding with constant velocity function block.

An Often Overlooked Aspect of Manufacturing and Packaging Systems Integration

Presented at the WBF
North American Conference,
March 5–8, 2006, by

Garvan McFeeley
Chief Information Officer
garvan.mcfeeley@urmasys.com
UrmaSys
IDA Industrial Estate, Collooney,
County Sligo, Ireland

Shane Loughlin
Manufacturing Technologies Director
shane.loughlin@urmasys.com
UrmaSys
IDA Industrial Estate, Collooney,
County Sligo, Ireland

Abstract

This chapter demonstrates that leading Original Equipment Manufacturers (OEMs) of manufacturing and packaging equipment can utilize both the work in progress on ISA-88 part 5 for a modular approach to automation and the security guidelines as outlined in ISA-99. This allows the OEMs to continue to earn continuous revenues over the life of the equipment by providing value-added services that increase the availability of their equipment to their customers. This concept is reinforced with real-world examples and quantified with data gathered from a project consisting of a fully automated manufacturing and packaging line for a world-class biomedical device manufacturer.

Introduction

This chapter is based on a project that consists of the implementation of four fully automated manufacturing and packaging lines over a 3-year period for a world-class biomedical device manufacturer. The main measure for success on this project is that each line was required to achieve an Overall Equipment Effectiveness (OEE) in excess of 90%. Other characteristics that added to the complexity of the project included 21 CFR part 11 compliance and the incorporation of electronic signatures.

In order to provide the highest chance of success, existing equipment was analyzed. The performance and quality of the process was under control, but the availability of the equipment was the main source of variability. Sources of downtime, which affected availability, were classified as either high frequency short duration or low frequency long duration stoppages.

In the following sections, both of the previously mentioned sources of downtime are explored, and the strategies that were implemented in order to minimize their impact are outlined in detail. The chapter goes on to explore the business models utilized by manufacturing OEMs and the unique business opportunity that exists whereby manufacturing OEMs can assist the end users to dramatically increase equipment availability.

High Frequency Short Duration Stoppages

If the concept of availability is to be introduced in a comprehensive fashion into a manufacturing facility, it must be kept in mind that the system should be easy to understand for all personnel in the manufacturing facility, the technology providers, and the OEMs. These parties should be able to converse clearly and concisely with minimum confusion.

The PackML state model[1] provides an ideal starting point because it includes a common terminology that all personnel in the organization, including operators, supervisors, engineers, and managers, can easily understand. This, combined with the fact that it may be incorporated into the ISA-88 part 5 standard in the near future, ensures that it is not a custom solution. By expanding the state model to include a few more states such as Starved and Blocked, a deep insight as to the machine's availability can be achieved.

With a machine-centric software approach to applying the PackML state model, as depicted by Figure 13.1, the model is transposed to the controller. Using a high-level language or Human Machine Interface (HMI) software at this point enables the data to be retrieved with Object Linking and Embedding for Process Control (OPC) and logged to a database. The resulting data can then be made available to any computer on the network via Web pages.

The machine-centric software approach provides many advantages, such as the following:

- The OEM has the most intimate knowledge of the machine and as such can create a state model that facilitates root cause analysis on their particular equipment.

- The application only needs to be created once. Thus the OEM can spread out the cost over all subsequent machines.

- The machine is "publishing" its data to industry standard interfaces. The cost of interfacing this data to other IT systems by the end users' IT department is drastically reduced.

- The machine data is available to be viewed by all personnel in the organization with the correct access rights without any licensing costs.

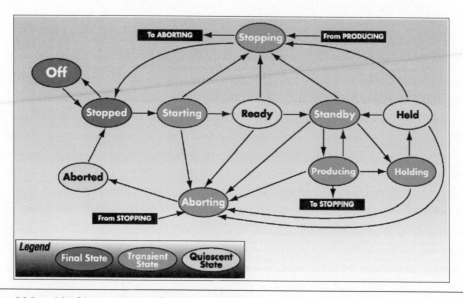

Figure 13.1.   Machine-centric software approach.

PackML State Analysis

In order to ensure that a consistent approach was utilized to maximize the availability of the equipment in the manufacturing facility, the following guidelines were utilized:

- Keep the approach simple and non-domain specific
- Decide on a few Key Performance Indicators (KPIs)
  - Use counters, PackML states (subdivided into machine issues and product flow issues), and alarms
- Adopt a Pareto Analysis, drill-down approach
  - Analyze counters
  - Look for periods of low production
  - Analyze PackML states for root cause analysis
  - Determine machine issues (e.g., faults, stops, and restarts; see Fig. 13.2)
  - Determine product flow issues (e.g., Blocked and Starved; see Fig. 13.3)
  - Analyze the alarms if a lot of time is spent in the Aborted state
  - Reduce the high frequency short duration alarms
  - Reduce the low frequency long duration alarms

Time Domain Analysis

Utilizing pivot tables allowed the manufacturing facility to leverage all its resources on one data set. The pivot table is a commonly used tool in both the financial and managerial domain. It allows the user to see different trends and facilitates both the summarizing of data to determine totals or the expansion of data to find the detail. It was found that there was virtually no training requirement for the managerial and administration level to use this form of analysis.

Production personnel, on the other hand, quickly identified how well they were performing on an hourly basis against their "competition" on other shifts, while engineering could focus on both the product flow issues and machine faults and restarts.

The elimination of auto scaling on the Y axis as depicted in Figure 13.2 and Figure 13.3 showed that all users were operating to the same metrics and that resources were not being misallocated.

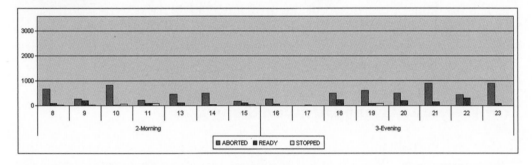

NON PRODUCING STATES - FAULTS & RESTARTS				
MACHINE ▾	DATE ▾			
	Apr 28 2004 Wednesday			
	STATE ▾			
	⊞ ABORTED	⊞ READY	⊞ STOPPED	⊞ Grand Total
SHIFT ▾ HOUR ▾	Sum of SECONDS	Sum of SECONDS	Sum of SECONDS	Sum of SECONDS
⊟ 2-Morning ⊞ 8	668	77	27	772
⊞ 9	253	201	14	468
⊞ 10	827	30	57	914
⊞ 11	224	90	97	411
⊞ 13	455	111	0	566
⊞ 14	512	40	0	552
⊞ 15	168	116	33	317
⊞ Total	3107	665	228	4000

Figure 13.2.   Non-producing states due to equipment.

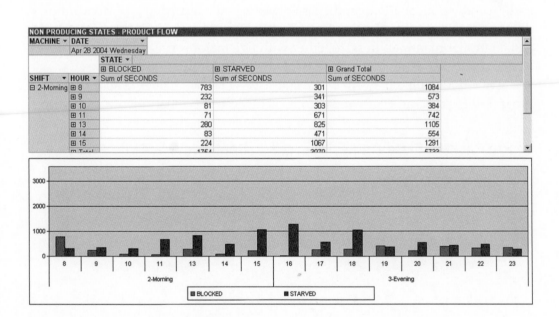

NON PRODUCING STATES - PRODUCT FLOW			
MACHINE ▾	DATE ▾		
	Apr 28 2004 Wednesday		
	STATE ▾		
	⊞ BLOCKED	⊞ STARVED	⊞ Grand Total
SHIFT ▾ HOUR ▾	Sum of SECONDS	Sum of SECONDS	Sum of SECONDS
⊟ 2-Morning ⊞ 8	783	301	1084
⊞ 9	232	341	573
⊞ 10	81	303	384
⊞ 11	71	671	742
⊞ 13	280	825	1105
⊞ 14	83	471	554
⊞ 15	224	1067	1291
⊞ Total	1754	3979	5733

Figure 13.3.   Non-producing states due to product flow.

The Business Case

The cost of developing and installing the machine-centric software approach was less than 5% of the cost of the overall machine. The resulting information allowed manufacturing and engineering personnel to increase the availability of the machine by 50% over a 4-week period, with minimal equipment investment. This increase in availability translated into an increase in production in excess of €1,000,000 annually on this one machine.

The resulting optimization of the machine availability is depicted in Figure 13.4. It should be clear from Figure 13.4 that even though considerable advances have been made, enormous opportunity remains. Thus it should be kept in mind that this is a continuous improvement process.

Low Frequency Long Duration Stoppages

The first step in analyzing the manufacturing facilities performance in responding to the low frequency long duration stoppages was to evaluate typical response times as a skills matrix, as outlined in Figures 13.5 and 13.6. Whereas the response times were reasonable, the complexity of the machines required detailed domain expertise for solving all but the simplest of problems. This is beyond the scope of engineers and technicians at an end user's manufacturing facility.

The variance in the control platforms as supplied by the manufacturing OEMs to this production facility over the last 20 years had resulted in an excess of fifteen different Programmable Logic Controller (PLC) models from eight different PLC and controller manufacturers, together with five major servo motion control platforms, connected using in excess of twelve different industrial networks.

Even in the unlikely event of the end user's engineers having all the required programming tools and software to connect to these previously mentioned devices,

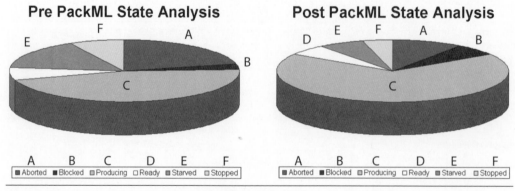

Figure 13.4.   Percentage of time in PackML states: Pre- and post-PackML state analysis.

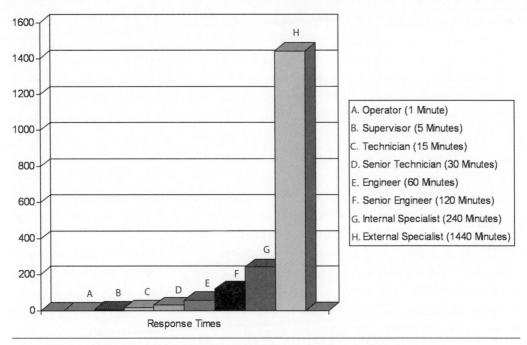

Figure 13.5. Typical response times to equipment stoppages displayed in minutes.

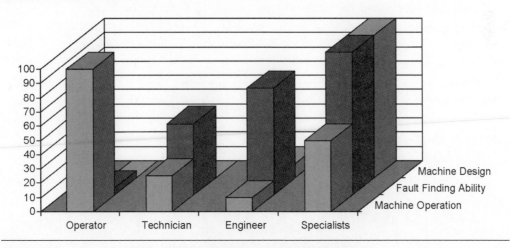

Figure 13.6. Typical skills matrix expressed in percent proficiency.

the training overhead of ensuring that these engineers were competent on this enormous variety of platforms would be prohibitive.

The real potential for eliminating these stoppages on new machinery could be found in ensuring that OEMs designed their systems in such a way that the operator or technician could be connected virtually immediately with the specialist in a manner outlined by Figure 13.7.

In order to ensure that providing this type of connectivity between OEM and end user did not introduce any unnecessary risks to the end user's manufacturing facility, the excellent work done by ISA-99[2,3] was leveraged on this project, and the resulting designs are described in the following sections.

ISA to the Rescue

Convincing end user IT departments to allow remote access to their infrastructure is a mammoth task and should not be underestimated. Any connection to the Internet is extremely difficult to justify from an IT security perspective. The ISA standards and associated papers provide an invaluable resource to ensure that a balanced, well-informed approach leveraging other end users' previous experiences is adopted.

The approach that was taken on this project and the relevant papers and the standards that were used were as follows:

- Perform general background research and basic education
    - *The Myths and Facts behind Cyber Security Risks for Industrial Control Systems*[4]

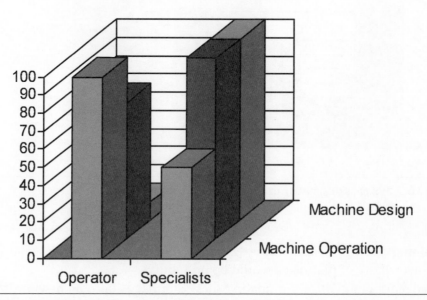

Figure 13.7.　Ideal fault finding solution expressed as skill set and required proficiency.

- *Current Status of Technical and Regulatory Issues Concerning Cyber Security of Control Systems*[5]

■ Evaluate the risks to the Manufacturing Control Network (MCN)

- *ANSI/ISA TR99.00.02 Integrating Electronic Security into the Manufacturing and Control System Environment*[2]

■ Determine the security tools that can be utilized to minimize the risks

- *ANSI/ISA TR99.00.01 Security Technologies for Manufacturing and Control Systems: Integrating Electronic Security into the Manufacturing and Control System Environment*[3]

■ Optimize the MCN design

General Background Research

Works by Byres and Hoffman[4] and Weiss[5] provide a valuable insight to the vastly different approaches to security taken by IT and manufacturing personnel. The main findings from this background research were as follows:

■ There is an accountability gap in many organizations.

- Awareness is still limited at the executive and operations levels. In addition, the CIO who has the responsibility for cyber security often has no accountability for control systems. Corporate culture has minimal accountability for cyber security of control systems. Very often, IT security or corporate operations is not aware of what networking and remote access has been implemented in the field.[4]

■ There is some statistical data on security incidents, but companies need to be more proactive in reporting attacks so that the overall bank of knowledge improves and is kept current.

- The British Columbia Institute of Technology Internet Engineering Lab (BCIT/IEL) maintains an industrial cyber security incident database that tracks incidents involving process control systems in all sectors of manufacturing. While most companies are reluctant to report cyber attacks or even internal accidents, there are now enough events to allow some basic statistical analysis of the data.[5]

■ It is very rare for attacks to come from the Internet.

- If attacks do occur from outside the plant floor, the infiltration rarely occurs directly from the Internet. Instead it typically is via backdoor connections such as desktop modems, wireless networks, laptop computers, or trusted vendor connections.[5]

- A company's own employees represent the biggest threat.
  - Employees caused over 50% of the recorded attacks.[5]
  - A study by the FBI and the Computer Security Institute on Cybercrime released in 2000 found that insiders carried out 71% of security breaches. This is supported by the realization that persons with high technical skill and process knowledge pose the greatest threat to an organization.[6]

This information has changed drastically with the advent of the Stuxnet virus in 2010.

Evaluating the Risks

ISA-99.00.02[2] provides an ideal framework for evaluating the risks associated with the MCN.

Applying ISA-99.00.02

A brief overview of the steps involved with applying ISA-99.00.02 is as follows:

- Define risk goals
- Assess and define the existing system
  - Form a cross-functional team
  - Perform pre-risk analysis activities
  - Perform a screening of inventory to identify and characterize manufacturing and control assets
  - Develop a network diagram
  - Update the screening inventory
  - Make a preliminary assessment of overall vulnerability
- Conduct risk assessment and gap analysis
- Develop a procedure and build countermeasures
- Define a component test plan

- Test countermeasures
- Define an integration test plan
- Perform a preinstallation integration test
- Define a system validation test plan
- Perform validation tests on the installed system
- Finalize operational security measures
- Perform routine security reporting and analysis
- Perform periodic audit and compliance measures
- Reevaluate security countermeasures

Initial MCN Network Diagram

Figure 13.8 outlines the initial MCN network diagram that was created by analyzing the proposed architecture as supplied by each OEM. This MCN network diagram was expanded to include the technology providers. The role of the technology provider is one that is often overlooked by many end users. Many machines that are currently supplied by OEMs are extremely complex, and very few OEMs have the in-house resources to completely support such a system, so they rely on the technology providers to provide additional support. This represents a serious security breach, due to the fact that the end users may not be aware of whom the OEMs are allowing to have access to the MCN.

From an IT perspective it may seem extremely wasteful to have eleven modems on one line as outlined in Figure 13.8, but normally the OEMs install the modems as part of their supply. Because each OEM may have a slightly different modem configuration, they are normally unwilling to consider other options. The cost of modems and the associated cost installation of an analog telecom line to the point of use were estimated at €1500 per point or €16,500 per line as outlined in Figure 13.8.

A major technical restriction on modems is their speed. All telephone line modems are limited to 56KB. The production equipment installed on this project will have complex motion control and many will have vision systems. Also the typical project size for the PLCs on this line will be in the 1MB to 3MB range, so it is not practical to provide support via modem. In practical terms a minimum requirement for the support of this machinery is network access with at least 1MB bandwidth.

The use of engineering laptops as programming tools by the end users' employees represents an extremely vulnerable situation from the end users' perspective, when Byres and Hoffman's[4] and Weiss's[5] work is considered.

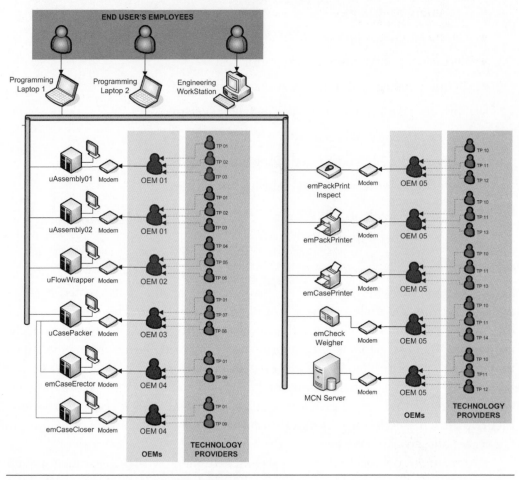

Figure 13.8.   Initial MCN network diagram.

Determine the Threat Probability Ratings

By applying the threat probability rating outlined in Figure 13.9 to the initial MCN network diagram in Figure 13.8, the following threat probabilities were classified:

- Category A (very likely)
  - 5 × OEMs
  - 14 × technology providers
  - 3 × programming terminals
- Category C (not likely)
  - 1 × integrated MCN

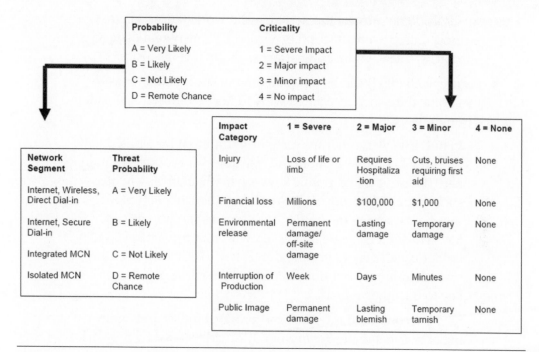

Figure 13.9. Threat probability and impact criticality rating.[2]

Thus the use of these modems introduces an enormous number of very likely threat probabilities to the MCN, while not delivering the required bandwidth and performance to allow the OEMs and technology providers to adequately support the supplied systems. Programming terminals, on the other hand, create substantial exposure to attacks from insiders and must be eliminated. The MCN should be isolated from the end user's Local Area Network (LAN) to prevent exposure from insiders also.

Determine Suitable Security Tools

Using ISA-99.00.01,[3] it was possible to identify suitable security tools to minimize the risks associated with access (both internal and remote) to the MCN. These tools are as follows:

- Virtual Private Networks (VPNs) are the preferred solution from an IT perspective because they are the industry norm, but they are not ideal from an MCN perspective because they allow unrestricted access to the LAN and do not offer any protection against the most frequent offenders—insiders.

- VPNs do not protect the network and workstations against most data-driven attacks (i.e., viruses), some denial-of-service attacks, social engineering attacks, and malicious insiders.[3]

- Secure Shell (SSH) can be implemented in such a manner as to overcome the issues associated with VPNs by routing the insiders through the SSH servers.

  - In public key cryptography, a pair of different but related keys typically known as a public-private key pair replaces a single key. The private and public keys are mathematically related such that a public key can be used by others to encrypt messages to be sent to the holder of the corresponding private key, which then can be decrypted with that private key. A key holder usually circulates the public key to other users in the same community but does not reveal the corresponding private key to the other users.[3]

  - SSH is a command interface and protocol for securely gaining access to a remote computer. It is widely used by network administrators to remotely control Web and other types of servers. Typically SSH is deployed as a secure alternative to the telnet application. However SSH also has the ability to do port forwarding, which allows it to be used in all three deployments listed previously. SSH is included in the majority of UNIX distributions on the market and is typically added to other platforms through a third-party package.[3]

  - There are no known security weaknesses with the dominant public key encryption algorithms.[3]

- Firewalls can be used to isolate the MCN, but they must be augmented with an Intrusion Detection System (IDS), and all access, both remove and insider, must be routed through the SSH servers.

  - While the firewall is the lock on the door to the process network, it is not the burglar alarm. You need some method of monitoring traffic and identifying malicious activity on the network. The tool to achieve this is known as an IDS and can range from a simple scan detector to a heuristic engine that profiles user behavior or a system that takes explicit action against the suspected intruder. In the process world, traffic patterns tend to be very consistent, so even simple traffic matrices that show who is talking to whom can be a big help.[5]

❑ The layered security model is very strong if it is implemented without exceptions. Unfortunately, we all know there will be exceptions. For example, a control vendor may need to connect to a PLC via a modem to offer technical support. As tempting as it might sound, banning non-standard connections outright is not usually feasible since the primary goal is continued production, not ease of security. What is needed is a system that can ensure that exceptions are logged and handled by means other than the standard firewall access. For example, a configuration policy and tracking system of all modem connections might be a first step. A more advanced solution might be to set up a secured remote access server attached to the firewall as a common dial-in point for all vendors.[5]

Minimizing the Risks

By applying the security tools that were identified to the MCN, an optimized MCN as outlined in Figure 13.10 was created. The following high-risk items were eliminated from the MCN:

- Modems
- Programming laptops
- Engineering workstations
- Raw connections to the Web

Every access point has a firewall and IDS as standard, which includes the following:

- Corporate firewall for Web access
- SSH server firewalls
- MCN server firewall

The SSH servers provide the following advantages:

- Two servers are provided for redundancy.
- All access to the MCN with programming applications are through the SSH servers. This includes access for the following users:
  - ❑ End user engineers

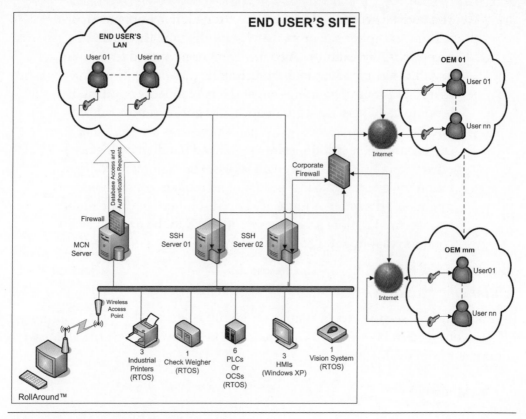

Figure 13.10. Optimized MCN network diagram.

- ❏ OEM engineers
- ❏ Technology provider engineers
- ■ The SSH servers have IDS together with network monitoring applications, so that the overall health and status of the network can be monitored at all times.
- ■ The SSH servers only forward the ports on the IP addresses that have been assigned to the user. This ensures the following:
  - ❏ The users only have access to the port on the IP address of the devices that they must support
  - ❏ One user cannot get access to another user's data or applications, even if they exist on the same IP address

It is important to note that the MCN network diagram outlined in Figure 13.10 achieves a significantly lower threat probability on all risks that have been identified. From a bandwidth perspective, the MCN can support 100MB and can easily be upgraded to 1GB as required. The only potential bottleneck in the MCN network diagram outlined in Figure 13.10 is the OEM's and end user's Internet access speeds, but 100MB connections are now commonplace and are more than fit for our purpose.

The Business Case

The cost of developing and installing the MCN that facilitates remote access as outlined in Figure 13.10 was actually less than the cost of implementing modems on the four lines. Thus it was justified from a hardware perspective alone, but the true benefit of the MCN in Figure 13.10 can be evaluated when the cost to production of solving technical issues is evaluated. In the following sections, two separate incidents are compared—one without remote support from the technology provider and one with.

Incident One

In May 2005, during the commissioning of the integral print inspection system, a serious issue was discovered by the system engineer, which meant that the system could not operate at the required frequency. The domain expert identified the problem almost immediately, once the expert was given access to the system. Unfortunately, enormous delays as outlined in Figure 13.11 were experienced during this process.

Even though it would be safe to assume that the faultfinding process would be expedited if the equipment were in production as opposed to commissioning, some very stark issues became apparent:

- An extremely long time was spent evaluating the issues and providing support. Even though both the engineers in question were very experienced, the fact that they did not have an escalation structure in place meant that they had nowhere to turn when they could not debug the issue.

- Support engineers attempted to emulate the problem in their laboratory, but they did not have the necessary equipment available and this caused unnecessary delays.

- The expert simply was not available when the problem finally escalated.

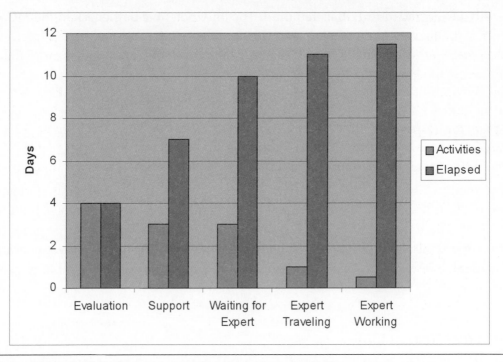

Figure 13.11. Fault finding process without remote connectivity.

- When the expert was finally available and traveled to the site, it only took him 4 hours to fix the problem.

If these issues occurred in a production environment, the cost of the issue would be multiplied by a factor of three for each working day that passed without a resolution, due to the fact that these plant operate on a 24-hour basis. Thus the actual problem, which took in excess of 11 working days to resolve, would cost 11 days × 24 hours × €2100 line revenues per hour, resulting in excess of €550,000.

Incident Two

In July 2005, there was another serious issue with the integral print inspection system, manifesting itself as an intermittent system crash. At this stage, the MCN was installed as shown in Figure 13.10. It facilitated immediate remote connectivity to the technology providers with no latency involved due to traveling or trying to emulate the problem. It also allowed the local engineer to ensure that only a small amount of time elapsed during the evaluation phase before he requested support. The support engineer escalated the issue very quickly to the expert, and on this particular occasion the expert had to escalate the issue to the system designer.

The end result was that the issue was solved in the same working day (Fig 13.12). In this instance, the expert was located in London, while the system designer was located in Austin, Texas. If both individuals had to travel to the site in a sequential manner, it would undoubtedly result in the loss of at least 1 week of production. Taking the initial line revenues per hour of €2100, this would result in a net loss of €382,500, without taking into account the direct engineering cost. The actual time lost of 10 hours as opposed to 168 hours represents savings in excess of €330,000 with this one incident alone.

Other interesting observations worth noting from this issue were the following:

- The fact that the system designer was located to the west of the expert meant that the working day could easily be extended without any work having to be carried out outside working hours by either person.

- The MCN in Figure 13.10 allows multiple connections to the system being diagnosed. At numerous times during this issue, all the engineers involved were on-line to the problem at the one time. The knowledge transfer and training process that this presents is not to be underestimated.

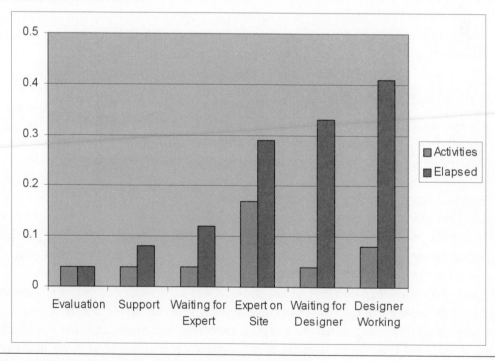

Figure 13.12. Fault finding process with remote connectivity.

- This method of connectivity facilitates a concurrent fault-finding process, as opposed to the traditional sequential model.

OEM Business Model

During the course of this project it became very apparent that even within this single end user's manufacturing facility, completely different OEM strategies had been adopted by the various OEMs. The product offering that was available from the IT OEMs was much more refined than the product offering that is available for manufacturing OEMs, and as such it provides much better equipment availability. Enormous opportunities exist for premium manufacturing OEMs to move the product offerings up the value stream by adopting similar business models to their IT counterparts.

IT Systems OEM Business Model

In this facility the IT department has engaged in a long-term, strategic, mutually beneficial, partnership-style arrangement with premium, world-class IT OEMs. This has resulted in an initial procurement price with a major support component that virtually guarantees very high availability. This commitment by the end user allows the IT OEMs to invest in substantial support infrastructure, which at a minimum provides a 24-hour 7-day help line with expert support on hand, remote connectivity and diagnostics, and next-day hardware replacement.

Manufacturing OEM Business Model

The engineering and manufacturing departments, on the other hand, had historically adopted an approach whereby the final selection of the OEM was decided by the procurement department, with the engineering and manufacturing department providing recommendations. In the absence of any other robust, quantifiable selection criteria, the default criterion used for the selection of manufacturing and packaging OEM equipment was price. It is not uncommon for the vendor to be selected on price alone by the end user's procurement department, irrespective of the engineering or manufacturing department's requests, because they cannot make a clearly defined argument for that particular vendor.

None of the OEMs on this project offered a support component as standard, but even if it were offered, it would be extremely unlikely to be procured because this equipment is categorized as a capital item and not a service. This approach means that OEMs have no support component in their revenue streams and can never be certain of getting the repeat business. Thus investment in structured modular coding techniques (as defined by the future ISA-88 part 5), support services, 24-hour 7-day help lines, improved documentation, and so on are simply not financially justified at present.

The current end user and manufacturing OEM relationship can be explained by Figure 13.13. It can be summarized that the OEM has no visibility to senior management except when it is escalated by production and engineering. Their services are not seen as being critical to the end user's business, and they are simply viewed as an engineering service provider who can be replaced or substituted at will.

Manufacturing OEMs have a unique opportunity to develop and refine their product offering so that they can reposition themselves in the marketplace over a very short time period. By leveraging the PackML state model together with a machine-centric software approach utilizing ISA-99 compliant remote connectivity and Web-based technologies, the manufacturing OEMs can drive enormous benefits in both their own organizations and in the manufacturing environments of their end users. This will allow them to move to the top of the value stream, as depicted by Figure 13.14, whereby senior management can immediately see the value-added services OEMs can offer. The business question for manufacturing OEMs is simply, Do you want to be selling to a person who has a budget or to the people who decide budgets?

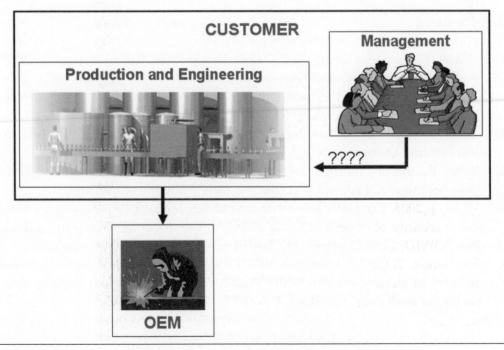

Figure 13.13. Current customer and manufacturing OEM relationship.

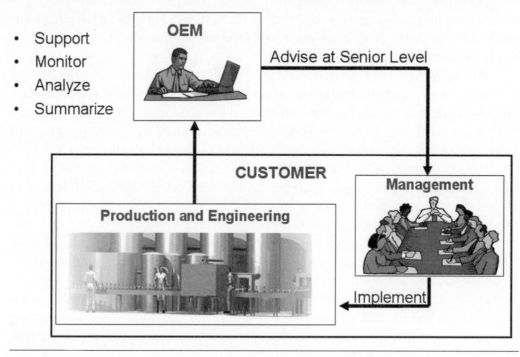

- Support
- Monitor
- Analyze
- Summarize

Figure 13.14. Potential customer and manufacturing OEM relationship.

References

1. OMAC Motion for Packaging Working Group PackML Subcommittee. 2002. Automatic mode machine states definition. OMAC V2.1. January 14, 2002.
2. ANSI/ISA TR99.00.02 2004. Integrating electronic security into the manufacturing and control systems environment. October 2004.
3. ANSI/ISA TR99.00.01 2004. Security technologies for manufacturing and control systems. October 2004.
4. Byres, E., and D. Hoffman. 2003. The myths and facts behind cyber security risks for industrial control systems. ISA technical paper. June 2003.
5. Weiss, J. 2003. Current status of technical and regulatory issues concerning cyber security of control systems. Paper presented at the 13th Annual Joint ISA POWID/EPRI Controls and Instrumentation Conference, June 2003.
6. Stephanou, T. (SANS Institute). 2001. Assessing and exploiting the internal security of an organization. http://rr.sans.org/audit/internal_sec.php (link no longer available).

Power Programming, ISA-88, and Packaging: Some of the Lessons Learned

Presented at the WBF
North American Conference,
May 15–18, 2005, by

Adam Maki
Senior Applications Engineer
Rockwell Automation
30 Airport Road
Suite 6
West Lebanon, NH 03784, USA

Introduction

Can the same standards for batch systems be used in the packaging world? Why would applying standards be worth the effort required to make it happen? How will this benefit the end user? Today's world of Original Equipment Manufacturers (OEMs) with their own proprietary code and structure will not be acceptable to end users that have seen a better method of programming machinery. The packaging world has gone from mechanical line shaft machines to electronic line shaft with integrated servo motion machines. With the increasing complexity of the control systems required to run these machines, we are compelled to provide a more structured approach to integrating these control systems. The industry is moving toward stricter requirements for data acquisition and standards that require dissimilar machines on a production line to produce the same data for efficiency information and product tracking.

The dream of plug-and-play plant machinery is a distant objective at best. Right now there are no real standard methods of defining the structure of motion control applications; let's work together to define them.

Transition from Mechanical to Mechatronic

Older packaging machines had a Start, Stop, and Jog button that would make the machine run. All actuators were tied to a line shaft that would turn pulleys and gearboxes to move end actuators. Coordination of these was done via mechanical methods (Fig. 14.1).

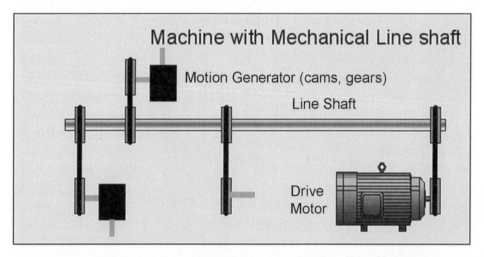

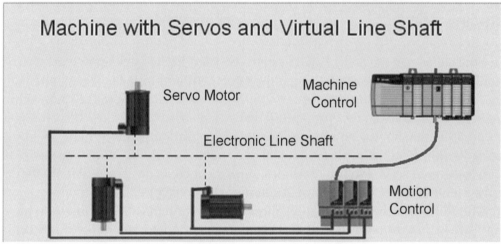

Figure 14.1.   From line shaft to servos.

Servo technology can replace the mechanical linkages and coordinate the control of these axes to work together with an electronic line shaft. This allows myriad benefits, including more precise positioning, faster changeover times, and so on. With the increasing complexity of the control system required to run these machines, we need to use a more structured approach to integrating these control systems.

The requirement to make products better, faster, and cheaper is increasingly in front of design engineers and programmers. How does one stay up to date with all the buzz words, guidelines, standards, and best practices that are being put into practice on a regular basis from different perspectives? Right now there are no real standard methods of defining the structure for different application types, but by working together we can define them. On joining the Make2Pack team, it became clear to me that there is a lot of work yet to be done to develop common definitions for the different ISA-88.01 models: physical, procedural, states, coordination, or what have you. The need to model them is critical to conveying the principles of structure.

I would like to share with you a brief look at the development of this methodology. As an applications engineer there are a number of things you need to do for any application. You will typically have your own code snippets that you have developed, understand, and use regularly. The Make2Pack team started out as a group of engineers who wanted to share best practices. Power programming got its start as an application note to define best practices for programming an integrated motion application. We started with a cartoner application note as the example. I came upon this state model for a machine called PackML and thought it could serve well as the behavioral model for a machine.

I worked to develop application layer routines to transition a machine program through these states, thinking that if we could stay within these states then it would be easy to share the expected normal behavior of the machine. I used this on a few applications and then bumped into another engineer who also used the PackML model for a machine. We talked and shared program codes and determined that we could actually understand and follow each other's code. This was an incredible epiphany! How often have you been able to look at someone else's code and readily understand it?

This became a focal point for discussing approaches to programming a machine with some of the best motion programmers on this planet. It became clear that there were divergent points of view even at the top of the game. This became a pursuit to find the best of the best. There were some good examples to look at, and most of the best programmers used some sort of state model that defined the behavior of the machine. The decision was to use the PackML state model and see if we could have it serve as the heart of the program, so that one could look at the model with only ten states and know how that machine was going to behave.

Using this state model as the foundation made code easier to write because there was a map to follow. The machine program will transition according to rules set forth by the state model employed. The PackML state model can be used to develop a functional specification, programming document, training document, user document, and Human Machine Interface (HMI).

When sharing the thoughts about this method of programming with machine programmers, their most common concern was that their code, which they had created and maintained, was going to be sent through a "pasteurizer." "Pasteurized, homogenized, and standardized" can also mean "indistinguishable, slow, and inflexible." This approach at first appeared daunting to some programmers, but once we discussed it further, it was generally regarded as a positive learning experience and they would consider how this could be applied to their application.

The problem with "best practices" is that someone always has a better one. We are defining some example best practices and giving you a template to work with. We are not trying to make all code the same for every application. The specific cam profiles or calculations you use will now have a convenient place to reside.

If people from the batch world took a look at what has been going on with motion control they would see that it has been undergoing similar development. Look at what has been accomplished in the batch world with regard to clearly defined standards and tools to model, create, and monitor systems. The future is wide open for motion control to follow the lead, and our challenge is to not reinvent the wheel. How did they do it? They started simple and built on the basics. We have seen the wizard! It is not one individual but a group of hard-working people working to define best practices and terminology. With this information it will be possible to communicate between different machine types and different hierarchies. Let's cross over the imaginary lines that divide our systems. This realization came as the result of my Make2Pack work. A lot of very valuable work has been done on the process side, but a lot of hands-on, good work has been done in the discrete world. This convergence of technology requires that we use the same terminology to find the common ground and both work toward the center.

Top Three Lessons Learned

1. Consistent structure makes navigation easier. Think of your favorite reference books. You can go to the table of contents and zero in on the chapter, section, or subsection you are interested in.

2. Try not to start from scratch. Instead, use a method:
   - Define your physical hardware and your procedures

- Define the exceptions
- Use the definitions

3. The demand for standards is increasing, especially for the following areas:

   - Validation that the developed system meets the user's requirements
   - Retraining requirements for new equipment
   - Genealogy requirements
   - Code reuse
   - Resource efficiency
   - Software and firmware designed for customer needs

End Users (The Production Plant)

End users are the ones that require these standards and guidelines. The production plant is required to fit all these unique machines and control systems together with bubble gum and duct tape. How nice would it be to have all the machines your company buys have a standard set of tags (names for pieces of data) and a standard behavior? Generally the customer needs are well defined:

- Access to critical efficiency information
- Maintainable code or a great support contract (see Chapter 13)
- Highest performance possible from the machine
- Low cost to integrate the machine into the facility
- A clean, intuitive HMI

The Controls Engineer

Leaner companies have fewer engineers, leading to more work per engineer. The resultant need for efficiency and better tools has never been more important. Add to that the increasing complexity of machine control systems and the need for standards integration, and you have a lot of homework to do.

Machine control programming was left for a top gun with an espresso dependency and a lot of field experience. He will likely have a personal library of programs and pull snippets from it for the entirety of his career. This is pretty typical, and most of the machine programmers that I contact are artists; they build elegant methods of programming the functionality their customers have been asking for.

Good structure has helped bring some organization into the programs of servo engineers, and this same structure can be applied to other applications. This can be done by defining the basic requirements and then looking at the exceptions and trying to make the programs behave according to a predefined behavioral state model.

Where does one become proficient with the latest technology? It can be in many places, including college, industry work, software training seminars, standards organization seminars, hardware design seminars, or the school of hard knocks.

Early in my work the goal was to utilize the PackML state model as a behavioral model for the machine. What we found on this pursuit was that we were indeed reinventing the wheel in some respects. It seemed clear at the onset of this project that there certainly must be a better way to write programs, but maybe what we really discovered is that there are different ways that may not necessarily be better. Other engineers within my company and outside my company were implementing methods similar to what we were doing. It is only upon further investigation that we realize this. Again, we were back in the school of hard knocks (where you had to make mistakes to gain experience). Technology is charging ahead, and machine complexity is increasing, and so are genealogy requirements, new software tools, standards, and customer requirements. So do we discard known methods to start from scratch with new software and new tools, or do we use old methods and adapt? The suggested answer is both: use new software tools and standards to help build upon your experience.

OEM

The OEMs with proprietary code and structure may not be acceptable to end users that have applied a standardized method of programming machinery. The packaging world has gone from line shaft to some servos to integrated servo motion machines. With the increasing complexity of the control system required to run these machines, technology providers are compelled to provide a more structured approach to integrating these control systems. Today's servo systems have copious data that can be produced for the user. It is not just a case of "get a bigger hard drive to hold the information." It must be managed to save the data we need and discard the data we do not. Really this boils down to good data structures and how to set up the data so that it is useful to the end user of the machine.

Some machine OEMs have this down to a science. They deliver a complete functional specification, operation manuals, parts lists, and so on. However, the customer always has to "test drive" and figure out how this particular machine behaves (even though they may have another one by the same OEM in the plant). A 2-week, on-site training program may be provided "free of charge." This type of individual masterpiece work is not what end users require. Two weeks of training is expensive.

The need for plant metrics will drive the OEMs to develop more standardized methods of structuring their code to be accepted by the end users or be relegated to subpar applications. An important part of gaining wider acceptance for your machine is being able to convey what the performance is going to be and to package data for the customer to integrate your machine into their system.

Controls Vendor

It is the duty of the controls vendor to listen to the customers and deliver tools that help implement guidelines, standards, and best practices. It will be the vendor that does this well that gets rewarded with customer dollars. Some of their biggest customers are internal customers—their own engineers—who require a better method of solving a problem or meeting the end customer's needs.

ISA-88 Standards

Coming at ISA-88 from a discrete packaging perspective, it seemed that the standard was cast in stone and that's that. This is not the case. A lot of good work has been done, but there is a need for more work and definition. I would like to briefly share with you some of the highlights of what I found most valuable from ISA-88, which could be applied to discrete machines. The definitions for the physical model are strong and can apply to discrete machines without a lot of mind bending.

The concepts behind the procedural model make clear sense, and the collapsibility of them makes them even more useful. Throughout my work with this model I have created a lot of scrap paper and gone back to the drawing board many times.

The definitions are important and might need to be sharpened. There are some deficiencies because they are meant to be widely applicable.

The ISA-88.01 model shown in Figure 14.2 brings together the different elements that are required for a system. This approach to mapping out hardware and software makes sense. It seems as if we are just on the verge of understanding a model or term when a different diagram or definition trips us up. From a machine programming point of view, it is important to see the interaction at the lower

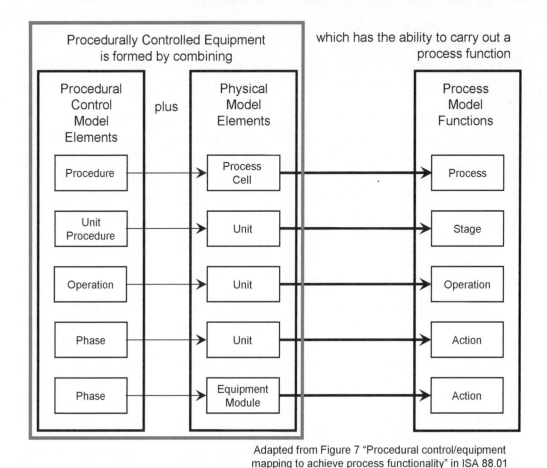

Adapted from Figure 7 "Procedural control/equipment mapping to achieve process functionality" in ISA 88.01

Figure 14.2.   Combining control and equipment to provide process functionality.

levels of control. Arguably this is an exercise for the reader, but there are a lot of great examples we can follow and learn from. If it provides the machine programmer with a more standardized method of structuring their program, then this will be a worthwhile effort. Thus the Make2Pack committee sees the value in it. If we cannot demonstrate the value of doing this clearly and openly, the programmers will not invest their time and effort just to adhere to another standard that some user requires.

The ISA 88.01 unit procedure model shown in Figure 14.3 is very useful, but I would suggest that for a machine we will collapse the equipment operation, so it would be

Equipment unit procedure → equipment phase.

The thought behind this is that most discrete machines are more basic and control oriented, and historically little regard is given to the overall procedural behavior because the machine generally will do only one procedure.

Physical Model

The ISA-88 physical model can be directly applied to a machine with some consideration to the layout and function of components. This machine—called the Vertical Form, Fill, and Seal (VFFS)—is going to be used as the example discrete packaging machine for this chapter. It forms a bag from a roll of film, fills it with junk food, and seals it into an individual bag. It can be broken down further as shown in the Figures 14.4 and 14.5.

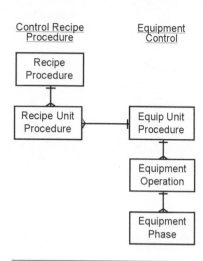

Figure 14.3. Separation of recipe and equipment control at the unit procedure.

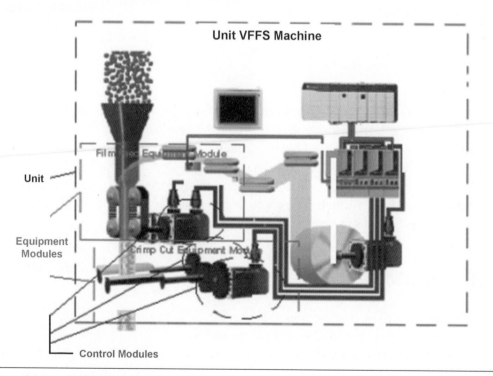

Figure 14.4.   VFFS machine.

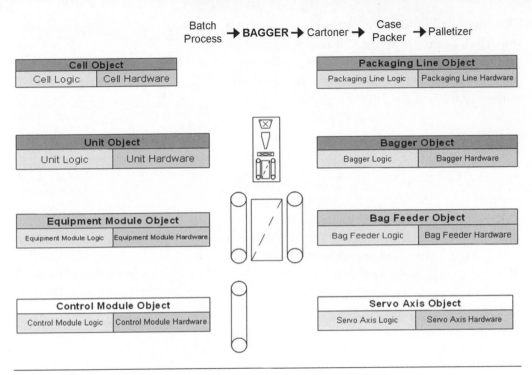

Figure 14.5.   Equipment decomposition.

PackML

It is clear there is some resistance to changing the way machines are programmed. By asking the question, "Does your machine behave according to this state model?" the first response is generally "No, my machine is very complex; there is a lot of synchronization that is required, and many things happen in parallel." Figure 10.4 illustrates how we can use the basic state model and describe the behavior of the machine. Thus the document begins to serve more as a functional specification, programming document, training document, and so on.

Behavioral Model

This section describes an example of using the state model to illustrate what must happen in a couple of states. When the power is turned on to the machine (in an Off to Stopped state) and the program is booting up, the following actions should be taken:

1. Establish communications
2. Wait for package glue to heat up

Next would be resetting and initialization (Starting state), initiated by a Start or Init command (Fig 10.4). When this occurs, the following actions should be taken (Fig. 14.6):

1. Clear all axis faults
2. Clear out products in a specific sequence (e.g., one conveyor and then the next in case of jamming)
3. Home all axes (e.g., synchronize the line shaft)
4. Enable electronic gearing

Figure 14.6. Components of the Starting state.

By working through all the states and defining the actions required within each state and the transition required to leave that state, one can clearly and concisely define the behavior of the machine. If we took our behavioral model and put it up on the HMI, the operator of the machine could see what the machine or unit is meant to do in each state, and what it is doing now.

Pack Tags

It is often troublesome to determine which tag names (Fig.14.7) we should produce on a per-machine basis and how to discern machine efficiency. With a clear set of tags, the same raw data is produced for all machines. Armed with the raw information in a consistent format, the Manufacturing Execution System (MES) and Enterprise Resource Planning (ERP) designers will be able to extract efficiency information and produce reports. Now that our machine is behaving according to the behavioral state model, by showing how long the machine has been in Stopped, Producing, or Aborted one can quickly surmise what the performance is. Deeper yet would be a reason code for why it transitioned into Aborting. A reason index could point to the fact that, say, guard door 5 opened.

Spec #	Prefix	Tag name	Tag descriptor	Data Type
3.72	PML_	Mode_Time	Time in current mode	Time
3.73	PML_	Cum_Time_Modes	Cumulative time in all modes	Time
3.75	PML_	State_Time	State time	Time
3.76	PML_	Cum_Time_States	Cumulative time in all states	Time
3.77	PML_	Seq_Number	Sequence number	Integer
3.78	PML_	Reason_Code	Reason code	Integer
3.79	PML_	Reason_Index	Reason index	Integer

Figure 14.7.   Partial table of names for machine parameters.

The Next-Generation Servo Controls Engineer

Let's take a look at next-generation machine builders. What if we begin to employ behavioral models that would define the inputs and outputs and how to get from here to there without having to know all about the code?

This may sound like a radical departure for machine OEMs, but it has been proven that they are smart people and will consider a new methodology. This is a different way to look at the same problem. Many of the discussions with OEMs have been something like "let's define the behavioral model for your machine and put it into a brief functional specification. Then we can decide what should happen in each state."

Program Layout

The state programming approach has been used for program layout for years, but the concerted effort to use a standard method has not. This methodology attempts to give the uninitiated user some "warm-ups" on state programming and get them thinking of their application not as "something completely different" but as a machine that can be commanded to operate in accordance with a state model that could be applied to a broad range of machines, lines, or control systems. This will become more useful when creating machines that interface with batch systems or other dissimilar control environments. The main goal is not to make all the machines the same but to make the commands and status feedback the same so the individual machines can communicate upstream and downstream.

Figure 14.8 shows a detailed version of Figure 10.4, with things to do in each state and the transition conditions to go to other states. The map is there—all that remains are the details.

Program Design Methodology

The following is an example of a program design methodology:

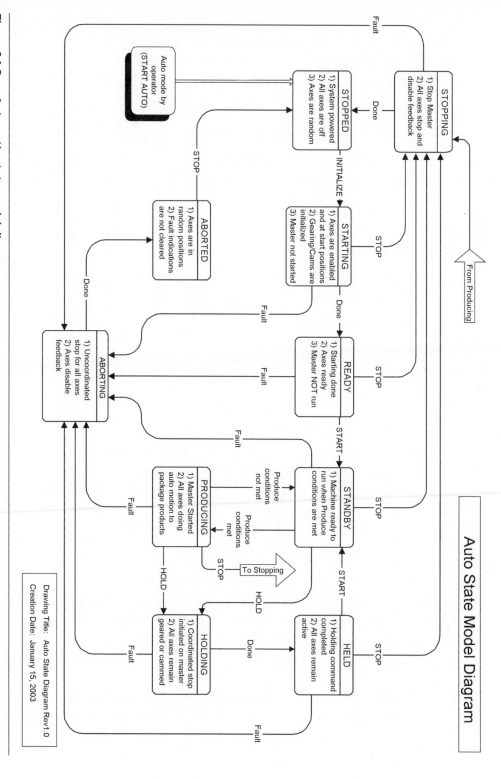

Figure 14.8. Automatic state model diagram.

- Define the physical model
  - Break the machine (which is an ISA-88 unit) into Equipment Modules (EMs) and Control Modules (CMs)
- Write the functional specification
  - Use the state model to create a quasi-functional specification
- Define the equipment procedural model
  - Use a program structure example
  - Do the normal machine operation first
  - Define the exception handling in as much detail as a risk analysis shows necessary
- Populate the program
  - Start with a good structure
  - Add CM, EM
  - Debug the program
  - Update the documentation
- Repeat as necessary

Define It

One thing that we have learned from this work is to set out a definition. Start out with a machine and define the physical model first; then look at the program and see how it should be structured. If there is some sort of linkage between the physical models, there is some structure and definition.

Conclusion

We are getting closer to the point of having the tools and guidelines we need to make a cohesive effort to develop standards that will likely make life as an engineer a little better. If you don't define it, you can't make it better. There must be a clear definition between the physical and procedural models for a machine or batch system.

A controller or some type of batch software is going to fill the requirement of controlling the procedural model. If a machine has a clear definition, the basic physical model may change and elements may be mechanically added or removed from the system without changing the procedural model.

Think about two programmers who write code for the same machine. One may write code that can produce more products at a better quality using the same mechanical equipment. The physical model didn't change—just the procedural model or program for the machine. This indicates that clear definitions and good structure are the backbone of any successful application.

Further Reading

ISA-88.01-1995. Batch control part 1: Models and terminology.

Fleming, D. W., and V. A. Pillai. 1998. *S88 implementation guide*. New York: McGraw-Hill.

OMAC PackML. http://www.omac.org.

Parshall, J., and L. Lamb. 1999. *Applying S88: Batch control from a user's perspective*. Research Triangle Park, NC: ISA.

Rockwell Automation Power Programming. http://www.ab.com/solutions/oem/powerprogramming.

Editor's Note

In the Further Reading section, please note that the works by Fleming and Pillai and Parshall and Lamb were written before ISA-88.01 had solidified. In particular, there were several interpretations of the properties of EMs at the time of these publications. Please see the 2006 revision of ISA-88.01-1995.

Using ISA-88 to Define a Compliant Packaging Environment

Presented at the WBF North American Conference, April 30–May 3, 2007, by

Moin Hussain
Manager, Process & Automation
mhussa1@mccus.jnj.com
McNeil Consumer Healthcare
7050 Camp Hill Road
Fort Washington, PA 19034, USA

Ron Williams
Senior Manufacturing Systems Consultant
ron_williams@entegreat.com
EnteGreat, Inc.
1900 International Drive
Birmingham, AL 35243, USA

Abstract

This chapter describes how a life-science manufacturer (the client) has applied the ISA-88 standard to define and sustain a compliant packaging environment. The client understood the strength of ISA-88 for batch processing but was unsure of how ISA-88 could support its FDA-regulated packaging operations. By engaging ISA-88 from the beginning of its new packaging program, the client came to realize the potential of ISA-88 to provide more than just an automated control systems definition. Using ISA-88 as the framework for the entire packaging line, operations (including change management, documentation, training, regulatory compliance, and configuration control) enabled the client to develop a comprehensive operations and governance program for the packaging line that spans all stakeholder functions and disciplines.

Business Drivers

History and recent events at the client's factory identified various operational issues (e.g., potential cross-contamination; potential use of incorrect versions of Standart Operation Procedures [SOPs], recipes, instructions, etc.; potential incorrect recording of non-conformance events) within its packaging environment, which could be resolved through the implementation of a compliant manufacturing environment. Opportunities to improve business performance through the automation of paper-based batch record processes as well as manual qualification processes for training records, materials, and equipment (to ensure that things are clean, calibrated, have good maintenance status, etc.) were also identified. Since the client was initiating a capital equipment project to deploy new blister packaging lines, it was determined that these lines would be designed and outfitted to support the client's shop floor integration manufacturing systems architecture and enable compliant manufacturing operations.

The client's objective with the shop floor integration program was to achieve a paradigm shift that eliminated sources of non-compliance within its manufacturing environment through the following measures:

- Automated qualifications (proper personnel, materials, equipment, recipe and instruction version, environmental conditions, etc.)

- Automated version control (the correct version of SOPs, work instructions, recipes, etc.)

- Enforced manufacturing workflow (enforcement of the workflow for automated sequencing and manual activities)

- Electronic batch records (batch record assembly, review, and dispositioning to support batch release by exception)

- Automated non-conformance management and corrective and preventive action (automated capture and management of non-conformance events and process alarms)

The client expected to improve regulatory compliance by eliminating sources of non-compliance through its shop floor integration program. The integration of enterprise-level systems directly to the production systems (e.g., packaging line equipment) supports improved regulatory performance by enabling closed-loop automated control of the process (as it downloads current recipe data and automatically collect process data). Business and supply chain performance will be improved through the replacement of manually produced batch records with electronic batch records. This will eliminate clerical mistakes and highlight exception conditions, thus streamlining the batch review and product release process. The ability to automatically capture shop floor data is also an important feature that supports the client's continuous improvement campaign.

The Vision

The client's vision for the shop floor integration program was to establish a set of new integrated capabilities that both control manufacturing processes and collect detailed information on these processes thereby enabling a platform for continuous improvement. The shop floor integration program developed a framework that aligned to the client's key business drivers (the client's 4 Cs):

- Compliance
  - Increased control of operators and equipment
  - 100% accuracy in reporting
- Cost
  - Visibility into waste streams
  - Tools to improve batch cycle time
- Commercialization
  - Closed loop between development, engineering, and production
  - Dramatic increase in prevalidation design and testing capacities
- Customer service
  - Increased precision in planning and manufacturing times and yields
  - Complete visibility into the full supply chain

Why ISA-88 Works for Packaging

The ISA-88 standard is established and proven as a comprehensive and reliable mechanism for defining flexible batch manufacturing processes. While packaging may not be considered a batch process, the ISA-88 standard is in fact well suited to support packaging operations. The ISA-88 standard provides a robust framework that accommodates the manual, semi-automated, and automated elements of packaging processes. The models shown in Figures 15.1 to 15.3 demonstrate how ISA-88 can be applied to blister packaging operations. The procedural control model shows how information is used within a packaging line (Fig. 15.1), while the physical model identifies the equipment that is used (Fig. 15.2), and the process model establishes how information can be combined with equipment to achieve the desired process functionality (Fig. 15.3).

The ISA-88 standard can also be applied as an overarching framework to define and sustain a compliant manufacturing environment (including compliant packaging) by serving as the basis for the following:

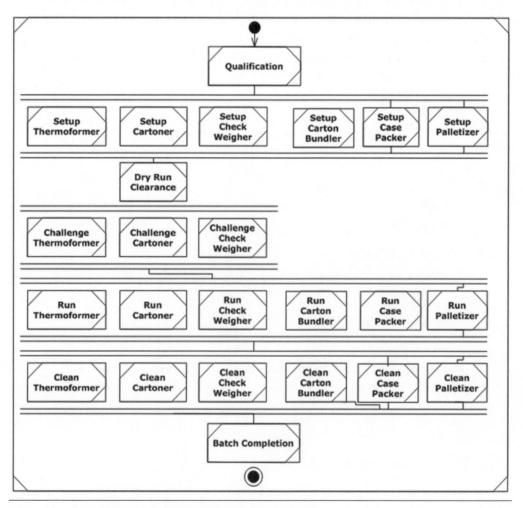

Figure 15.1. New blister ISA-88 procedural control model.

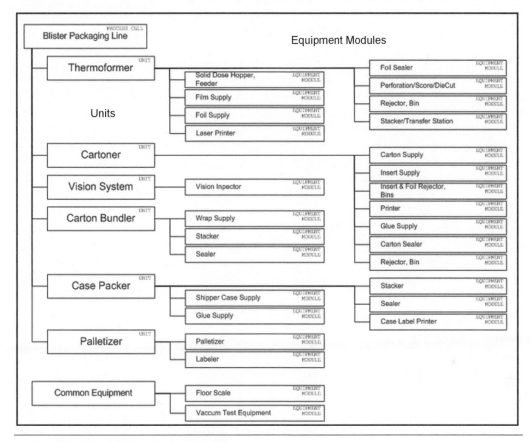

Figure 15.2. New blister physical model.

- Process definition (automated sequencing and manual activities)
- Master data definitions and recipe structures
- Integration with control systems
- Regulatory compliance
- Change management
- Documentation
- Training
- Governance

The ISA-88 physical, procedural, and process models were developed for the client's new blister packaging line to define the automation and control system behavior. These same ISA-88 models were also used as the framework for definition

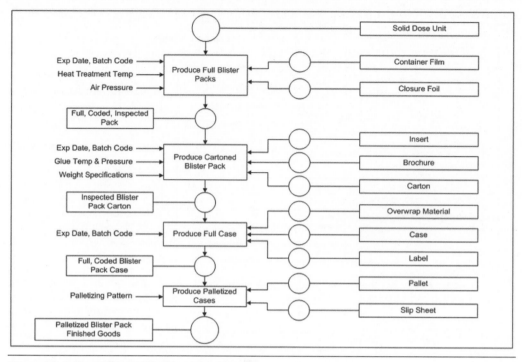

Figure 15.3. New blister ISA-88 process model.

and creation of the non-system-related and procedural elements of the compliant packaging environment: change management, documentation, training, and the ongoing governance. The ISA-88 models established an effective platform to integrate various organizational functions and disciplines to comprehensively define and sustain the compliant manufacturing environment.

Compliant Packaging

Compliant packaging is a manufacturing process designed and executed to eliminate the opportunity for non-compliant events to occur, thereby providing maximum assurance that all product generated by the line shall in fact be compliant with regulations and company policies. A large portion of the compliant packaging environment involves automation to support processes that were previously handled with manual exchanges of information (e.g., batch record data transcription and calculations, process and recipe setup on processing equipment, manual capture and transcription of non-conformance events, manual review of each field on a batch record, etc.). A comprehensive compliant packaging environment also requires structured procedural elements that support packaging operations (e.g., documentation, training, change management, and governance).

The systems-based automation portion of a compliant packaging environment engages systems integration from the enterprise level through supervisory execution and batch management to the control systems level. This approach has enabled the enterprise to achieve the required functionality in a structured, consistent, and comprehensive manner across all levels of its systems architecture. A critical aspect of this architecture is the integration of Enterprise Resource Planning (ERP) master data and recipe management with the packaging control systems. This integration is achieved through separate batch management functionality that coordinates phase-based execution for the overall systems architecture.

In particular, this batch management functionality handles the orchestration of recipe execution across the enterprise and control systems levels while also providing master data synchronization to avoid duplication of data and to ensure there is only "one version of the truth." The master data definitions, which are based on the ISA-88 physical, process, and procedural models, are established and maintained within the ERP system. This information is synchronized with other systems in the architecture, including the batch executive and data historian, which prevents unnecessary data transformations when information is exchanged between systems. These levels of integration enable a systems-based compliant packaging environment that leverages the master data resident within enterprise-level systems and assures that the current version of production instructions (e.g., Bill Of Materials [BOM], recipes) are engaged from the enterprise-level directly to the packaging control systems. Additionally, enterprise-level integration also provides access to the data necessary to support shop floor qualification activities (e.g., materials, personnel, equipment, and environmental conditions). This ISA-88 based overall shop floor integration architecture enables compliant packaging through the following:

- Enforced workflow
- Creation of electronic batch records (and elimination of clerical batch record errors)
- Exception-based batch record review
- Centralized recipe management
- Non-conformance management

Additionally, this system supports change control through a mix of automated processes as well as manually performed procedural control activities. Maintaining synchronization of the system and integrity of the data is essential within life science manufacturing. As such, it is imperative that all changes are procedurally controlled across both the enterprise- and execution-level systems.

Regulatory Compliance

The ISA-88 models can be applied to support a quality system inspection technique approach for "top-down" quality management. The centralized recipe management and qualification management capabilities supported with the client's new blister packaging lines are based on a standard ISA-88 structure that extends from the enterprise-level systems to the packaging control systems. The client's compliant manufacturing system also provides one system of records with one set of master data thereby enabling "one version of the truth."

The client's shop floor integration program for new blister packaging also enables the collection of shop floor data through both automated sequencing and manual data entry for the creation of electronic batch records in a manner compliant with 21 CFR Part 11 (electronic records and electronic signatures). The data is collected in a manner consistent with specific phase-based logic according to the ISA-88 model. The collection of data related to non-conformance processing events is also provided in a manner that highlights these exceptions within the batch record. For example, within the Run phase, various process actions are performed that may create or identify non-standard events such as the automated collection or manual entry of improper data that exceed established limits. When these events occur, the system presents an error message and records a description of the non-conforming event. Even as this non-conforming event is resolved, the batch record maintains a description of the non-conforming event, which is highlighted as an exception during the batch record review.

Change Management

The new blister packaging ISA-88 model was utilized to define change management requirements for the project. The roles and responsibilities for operating and supporting the new blister packaging line were established by mapping the ISA-88 procedural control model to the client's organization structure. Swim lane diagrams and the ISA-88 procedural control model were used to establish activities and responsibilities for each role and position involved with the new blister packaging line. This approach provided a comprehensive definition of the staffing and skills needed for operation and support of the packaging line and highlighted gaps or disconnects associated with potential staffing plans.

The new blister packaging line's change management program was based on preparing the workforce for the responsibilities of each individual role. The necessary competencies required to perform each role were determined along with the appropriate metrics and Key Performance Indicators (KPIs) needed to evaluate performance within each role. Based on the required competencies identified for each role, corresponding developmental plans were established. This approach, which could be traced back to the ISA-88 model, provided the framework to align the new blister roles and responsibilities with an overall change management program that encompassed communication, awareness, organization structure, culture, competencies, and training. As such, each role within the packaging team was mapped out using swim lane diagrams and the ISA-88 procedural control model to determine the necessary competencies and skills for each position, which in turn led to the creation of focused training plans. Additionally, the swim lane diagrams were also used to map KPIs for each critical phase of the operation to establish performance targets and objectives, thereby creating accountability for the activities as well as line-of-sight visibility from those KPIs to comparable business-level KPIs for that processing area.

Documentation

Documentation that defined the operation and support of the new blister packaging line (in the form of SOPs and work instructions) was organized and mapped using the ISA-88 procedural control model. These operating documents—which are important elements of all manufacturing environments—have additional significance within Current Good Manufacturing Practice (CGMP) manufacturing environments, as they establish the written procedures that are presented to the operating team in read-only electronic form for production and process control. As such, these documents are subject to regulatory review and must comply with applicable regulations. For the new blister packaging environment, these documents, which were created, modified, and maintained in electronic form, were organized in the following manner:

- *SOPs.* These documents define the overall process and macrolevel procedures within the new blister packaging environment (such as equipment qualification). The new blister packaging SOPs were organized to align with the procedures and unit procedures within the ISA-88 procedural control model (e.g., qualify materials, qualify personnel).

- *Work instructions*. These documents provide detailed instructions for specific operating activities within the new blister packaging line environment. The new blister packaging work instruction documents were organized to align with operations and phases of the ISA-88 procedural control model (e.g., qualify thermoformer, qualify vision system, qualify cartoner) within specific procedures or unit procedures and the corresponding overriding SOP documents (e.g., qualify equipment).

In general, the SOPs are product independent, whereas the work instructions may or may not be associated with a specific product or range of products. The ISA-88 model defines and maintains the integrity of the relationships between these documents. As the ISA-88 model is adjusted to reflect changes to the overall manufacturing process, impacts to the corresponding operational documentation can also be determined and managed.

Training

The ISA-88 model was utilized as a framework to help define the training content and program for the new blister packaging line. As is the case with the line's operational documentation, operational training within CGMP manufacturing environments is subject to regulatory review. Specifically, as stated in the FDA "Guidance for Industry, Quality Systems Approach to Pharmaceutical CGMP Regulations, September 2006," managers are expected to establish training programs that include the following for quality systems:

- Evaluation of training needs
- Provision of training to satisfy these needs
- Evaluation of effectiveness of training
- Documentation of training

When operating in a robust quality system environment, it is important for managers to verify that skills gained from training are implemented in day-to-day performance.

The ISA-88 model was used to drive an overall training program within the new blister packaging manufacturing environment in the following manner:

- *Training plan.* This plan defines the overall training program, training deliverables, training structure (i.e., partitioning of training content across roles and specific courses), and processes (e.g., courses, sessions, and course content and agendas) for the initial training rollout. The training plan was organized as a component of the overall change management plan, which has role and responsibility alignment with the ISA-88 procedural control model.

- *Training matrix.* The training matrix provides a detailed cross-reference (via the ISA-88 procedural control model) for the specific roles and functions with operational documentation (e.g., SOPs and work instructions) and training documentation and materials. As such, the training matrix creates alignment between the training content and the various training participants through alignment with the ISA-88 procedural control model to assure coverage and training on each operational activity.

- *Gap analysis report.* This report identifies all relevant operational documentation (e.g., SOP and work instructions) that will be engaged within the training program and links these documents to the ISA-88 procedural control model to identify the operations documents that require updates, should be retired, or are missing and need to be developed.

- *Training content and materials.* The content used for training is organized and developed in a modular fashion that corresponds to the ISA-88 procedural control model. Training content that is organized and built in this manner provides significant flexibility to support engagement of the content within various training sessions for specific audiences and roles while leveraging the same training content.

- *Training records.* Within CGMP manufacturing environments, maintaining detailed training records is critical to support both regulatory review of training programs as well as batch record review and dispositioning and batch release procedures. Qualification of personnel to perform various operational and quality activities requires access to appropriate records that demonstrate that these personnel have been properly trained. Therefore training records that document the training status of personnel involved in the production of a batch need to be verified as part of that batch's record review and product release procedures.

- *Training assessment.* To verify that the training of operational and quality personnel was effective, assessment tools in the form of examinations were included at the end of each training session. The results of these assessments were also recorded in the training records to provide verification that the appropriate skills were developed.

Governance

The governance model for the overall new blister compliant packaging environment is based on the ISA-88 model. As such, any proposed changes to the new blister packaging environment are evaluated in terms of their impact on the ISA-88 model. When appropriate changes are determined, the ISA-88 model serves as the governing framework to coordinate synchronized modification of all affected elements of the overall packaging environment (e.g., software, documentation, and training materials) in a comprehensive manner. Maintaining the integrity of the ISA-88 model is critical to sustaining the ongoing engagement of all new blister packaging line stakeholders.

Using ISA-88 as the overarching framework for governance of a compliant packaging environment enables a common definition language across all stakeholders, promotes cross-functional and cross-disciplinary ownership, and provides a mechanism for comprehensive evaluation of proposed changes. The governance model includes business rules that establish the related workflow and escalation associated with any proposed changes to determine the scope and magnitude of any suggested process changes. This evaluation provides for a comprehensive assessment and inventory of all associated modifications (e.g., documentation and training) that will be necessary to sustain overall system integrity.

Summary

For the client, its new blister packaging program was a valuable learning experience to understand the broad applicability of the ISA-88 standard. By engaging the ISA-88 standard from the beginning of the program in an unconstrained and open-minded fashion, the client leveraged the standard to the fullest extent possible. As such, the client was able to develop a logical and straightforward yet comprehensive method for defining, implementing, and sustaining its compliant manufacturing environment. The new blister packaging program clearly demonstrated the applicability and value of ISA-88 beyond its traditional role as a batch processing modeling methodology. In addition to demonstrating its capabilities as a toolset for defining packaging operations, the ISA-88 standard also proved to be an effective framework for mapping the related non-systems-oriented, procedural aspects of the client's compliant manufacturing environment, including documentation, training, regulatory compliance, change management, and governance. With this methodology and framework in place, the client is now positioned to expand the role of ISA-88 within existing operations and engage the standard in a similar comprehensive manner on future programs.

Software Modularity: A Powerful Tool for Packaging Automation

Presented at the WBF
North American Conference,
March 5–8, 2006, by

Tom Jensen
Senior Technology Evangelist
tom.jensen@elau.com
ELAU, Inc.
165 E. Commerce Drive
Schaumburg, IL 60173, USA

Leif Juergensen
End User Business Development Manager
leif.juergensen@elau.de
ELAU AG
Dillberg 12
97828 Marktheidenfeld, Germany

Introduction

Modular software design benefits packaging automation for the same reason that WBF recommends modularity in process automation. When implemented with ISA-88 (IEC 61512) principles in the IEC 61131-3 international standard automation programming languages, modularity becomes a powerful tool for both machine builders and users.

The cost is up-front development of programming best practices, templates, and software object (IEC function block) libraries. The benefits include documented engineering time savings of 70% to 80% by machine builders, better engineering collaboration between Original Equipment Manufacturers (OEMs) and specifiers, and ongoing reductions in users' total cost of ownership through reusability, consistency, maintainability, mechanical modularity, and faster execution speeds. In addition, validation savings of 40% have been documented in regulated environments.

This chapter will compare examples of typical control functions written in monolithic and modular formats, as well as best practices for combining the IEC 61131-3 languages and ISA-88 (PackML) constructs, leveraging other emerging technologies, and managing the logical design, process design, and deployment and architecture.

This chapter will also define the business imperatives and benefits driving software modularity, including achieving faster time to market, capturing operational efficiencies, and streamlining Management Execution Systems (MESs) and supply chain management implementations, maintenance outsourcing, and more.

The State of Machine Programming

Before we can tell the story about software modularity and efficiency, we must review a bit of the past. For many years and for many reasons, machine programming has been implemented by designing for each machine. This fact has several roots. The first reason is straightforward; the first machines were very simple. The control structure consisted of a Programmable Logic Controller (PLC) that was replacing relay logic (if you were lucky). The drafter creating the schematics for the machine builder usually drew the wiring of electrical hardware as a ladder diagram. Ladder diagrams were a clear and graphical way to show the relationships among simple physical objects (e.g., switches, relays, lights) as they were found on the machine. It was a logical and user-friendly offering for PLC providers to extend ladder programming to a technology wary public. By offering ladder language programming, the PLC providers of the day were changing the hardware, not the rules.

Furthermore, as programmers became comfortable with the easy adoption of ladder programming, they didn't feel the need to move to another language. This didn't mean that other languages weren't available; it meant that ladder programming was more than adequate for Generation 1 (G1) machines.

As machines became more complex (especially after the introduction of variable speed motors, programmable limit switches, temperature controls, and servos), machine programs grew larger to articulate the newfound sophistication of machines. It was no longer enough to "start a motor" or "check a switch;"

programmers needed to find ways to "home" motors, synchronize them, adjust travel distances, and have them recover lost positions. The programs created with ladder programming began to look less and less like the machine that was being controlled. Instead of the single line of code that showed the interaction between physical hardware objects (e.g., contact turns on relay), it took large parts of code to describe the actions of a single motor. The end result was machines that were highly complex with limited flexibility and difficult to modify, integrate, trouble-shoot, and own. On average, today's packaging machines contain multiple motors, temperature controls, programmable limit switches, load cells, and weigh scales. They are also responsible for machine performance data, product data collection, preventive maintenance features, and more. They also need to conform to more stringent government regulations.

The Language Problem

Ladder programming proved itself useful for describing massive amounts of Boolean expressions but not as useful for describing complex machines. It was commonly realized that other languages could be helpful in describing these more complex G2 machines. There were many languages available from the industry, but they all suffered from one of two conditions: First, the languages commonly offered on the market were either text based or graphical. Text-based languages in general were capable of describing complex objects with much less code, but they were not intuitive. This meant a programmer could actually create machine code that worked well and described how that machine should operate, but nobody without programming education would be able to work on the machine. Secondly, there were graphical languages. They were intuitive but lacked the depth to com-pletely fulfill the desire to have replication of the real world in code.

The IEC 61131-3 Proposition

In the mid nineties, a standard was developed in Europe to address the "single language" approach many technology providers had adopted. A suite of five languages had been adopted as a worldwide standard for programming PLCs, named IEC 61131-3. The standard provided for the nesting of one language inside another, structures, complex data types, libraries, templates, and more. It provided a way for programmers to develop code the way a programmer does, but then to display the code to a machine technician in a graphical way. The 61131 standard promised the ability to modularize machine software and to deliver truly reusable code that would do the following:

- Increase machine performance
- Reduce delivery time to market
- Improve serviceability
- Speed the validation process
- Reduce the total cost of ownership

Sounds Good . . . How Do I Begin?

G3 machines are far different than G1 machines. A G1 machine could be programmed as a monolith—one big block of code—because there was not much to them in terms of electrical controls. For example, a G1 tube filler may have had one main motor driving a Geneva gearbox, with all motion mechanically cammed to it. On the other hand, G3 machines are electrically complex and should be broken down into the smallest possible pieces (granularity). The smaller the pieces, the more reusable the code will be. A G3 tube filler consists of servo-controlled pumps, tracks, nozzles, and sealing bars. Also there are temperature controls, metal detection, and Human Machine Interfaces (HMIs) to consider.

Now that we have drawn a distinction between how things have worked in the past and where we would like to be in the future, we need a plan to get us there. The first consideration should be modeling our machine with granularity in mind. If we are programming a tube filler, we may have created a machine model that consists of the following mechanisms:

- Pump and manifold
- Tube loading, tube track and orientation
- Filling nozzles and seals
- Gates between the filler and the timing belt to the cartoner
- HMI and data collection
- Temperature control for the end seals

These mechanisms are all made from the following components:

- Servos

- Variable frequency drives
- Touch screens
- Input/Output (I/O)
- Heating elements

With these in mind, we now have enough of a map for our machine that we can begin programming. We should begin by making the smallest components first, keeping in mind that in some way operators and maintenance people are going to need access to them for setting up and operating the machine.

Let's start with making a piece of modular code for a servo motor. Once we model the servo in a function block (a software object as defined by IEC 61131-3) and store it in a library, we can drop it into any other code module where a servo is needed. At this point we need to decide which language would be the best, and the obvious choice would be the language "Structured Text" because of the mathematical demands as well as the complicated sequencing needed. Any of the five languages will do, but the goal is to reduce code. We will define a number of input and output variables to the function block we are writing and set out. All possible needs (e.g., operating modes, inputs, outputs) should be provided for at this point, no matter how "rarely" the feature may be used. Why? Because if we spend a small amount of time at the front end to take care of all needs, it will preclude us from revisiting and rewriting later. (If you make a good list, you will only have to go to the grocery store once!)

Once completed, the servo function block should be tested, documented, and locked. Whether we have one, two, or one hundred servos in this program, this is the only function block we need. We follow the same process for the other simple components listed with the servo and create a locked library, as is provided by IEC 61131-3. Even though we may have used different languages for the function blocks we've just made, we can call them into the next layer in whatever language we want. This means the servo object we made in Structured Text can be displayed as a ladder diagram object, creating a graphical representation of the machine that most technicians will be comfortable with.

Modules for Machine Variants

Now that our simple objects have been saved as a library, we can start programming the mechanisms that compose our machines. Again, we can pick a language that suits us. But there is one problem: the machine we are programming is one in a family of machines. The smallest operates on two tubes with one color and the largest on eight tubes and two colors. The process we used for the simple objects still holds true. We should write the code to accommodate the largest, most difficult machine (we will have to do it anyway), and make the code able to be configured by parameters to fit the smallest machine as well. When we do this, we will be creating a library of a few objects that can be placed into any tube filler our company makes, and we can commission the machine by setting parameters.

We have hit a milestone: by adopting a process for developing code (e.g., ISA-88, Make2Pack), we have just leveraged a practice that will improve our business in the following ways:

- The cost of engineering code will decrease
- The cost of developing new machines will decrease
- The integration of new features will become simpler
- The stability of our products will become better
- Delivery time will drop
- Documentation and validation will decrease

Programming the Top Level

Now that we have made simple components and mechanical modules, we need to tie them together in a framework. The consideration at this point should be to simplify line integration, improve connectivity to upper-level systems (e.g., MESs), and promote service and diagnostics. And of course, the machine should be simple to operate. This is where the unique code that makes this machine different from all others will be written. All lower layers of code are made of objects we've written for the machines our company makes (e.g., tube fillers, cartoners, case packers). But how unique are most machines? They all have the same basic composition. They need error checking, inputs, outputs, mode control, messaging, and standard tags for integration.

If we employ the language best suited to describe a structure to act as a virtual filing cabinet, we can greatly simplify the code at this level. We may not be able to put it into a library, but surely we can make it into a template. Then any service person, operator, or IT engineer will be able to get the information they need. Therefore we should use Sequential Function Charts (SFCs) in this case and again use a top-layer standard (e.g., PackML) to standardize the tags needed for all these features. By using SFC and PackML at the top layer and standard libraries for the components below, we have created a machine that has the following benefits:

- Reduced installation costs

- Reduced line integration costs

- Better overall equipment effectiveness and root cause analysis

- Simpler operation and troubleshooting

Software Modularity Put into Practice

The process of changing monolithic code to modular code has been gaining momentum over the last few years. A number of companies have gone far enough with the process to share results.

Douglas Machine, Inc.

At the Packaging Machinery Manufacturers' Institute (PMMI) Conference at PACK EXPO 2005, Joe Faust, Electrical Engineering Manager for Douglas Machine, shared his company's experience with the underlying software architecture, the initial cost to build a foundation of modular software using the IEC 61131-3 languages, and the resulting benefits.

This experience was eye opening, revealing the competitive advantage to Douglas as an early adopter. The methodology is simple. Machine builders can reduce engineering costs and increase consistent quality through the use of libraries of tested, reusable software modules that "plug" into a state model as prescribed jointly by the Organization for Machine Automation and Control (OMAC) Packaging and Make2Pack groups.

Combined program development, program testing, and machine testing with modular software took 28 to 34 days, compared to 50 to 65 days for a conventionally programmed or "monolithic" programs. Programming time was reduced by at least 50% (from 20 to 25 days down to 10 days), and program testing was reduced by approximately 80% (from 15 to 20 days down to just 3 or 4 days). The savings potential was clearly derived from better software engineering, as the machine testing component remained unaffected, at 15 to 20 days.

Faust cited shorter development time, standardized program structure, the ability to increase machine features, ease of machine upgrades, the ability to expand the breadth of machine offerings, and the ability to protect intellectual property as benefits to machine builders as well as users. Faust noted the up-front costs as well. First the modular foundation must be developed, and for the first modular machine, this took 100 days of software development and 30 days of software testing, in addition to normal machine testing.

Smurfit Kappa Herzberger

Kappa Herzberger, recently merged with Smurfit-Stone, specializes in high-speed packaging machinery for the beverage and food industries. The company applied IEC 61131-3 conforming modular programming techniques in a complete redesign of its product line, according to Michael Heise, Manager for Automation Technology.

Through the use of these modular software building blocks, Kappa has consequently reduced engineering time by 50%, while the new machines realized 75% higher cycle rates, 35% less commissioning time, and 60% faster format changeovers.

HAMBA Filltec (Now Part of OYSTAR)

HAMBA Filltec has applied no fewer than nineteen servo-automated functions for flexibility in plastic bottle, cup, and tub filling and sealing machinery. The container station uses servos for container gripping, automatic de-nesting, tub transport, and scissors operations. The metering unit uses servos for lifting and piston and valve rod actuation. The lidding station uses servos for lid gripping, application, and drum movement. At the sealing station, the sealing bridge, lift mechanism, support, and closure systems are all servo operated, as are leak testing, container ejection, and chain cycling functions.

Benefits of the modular mechatronic design include not only rapid pushbutton format and product changeovers but also convenient operation and maintenance. For example, with two lidding station magazines accessible from outside the machine, one can be reloaded while the other remains in operation. The upper section of the sealing station has been designed to pull out and to the side for easy cleaning, setting, and maintenance access. The systematic implementation of a new generation of diagnostics makes fault detection and troubleshooting easier than ever for operators.

A high-end automation solution is essential to control such a flexible machine, which requires up to thirty-four servo motors according to Uwe Gerasch, Development Manager for HAMBA. To do so, the company implemented a completely modular control software structure. The control system integrates motion and logic control in the same IEC-conforming program.

The control system also operates up to 1000 distributed digital inputs and 1000 digital outputs, plus 50 to 170 analog I/O points per machine. Instead of discrete hardware controllers, software loops control heating circuits and sterilization, while automated software cam adjustment controls numerous cam positioners throughout the machine. In addition to interactive diagnostic guidance from the touch panel, the operator also receives graphical assistance, fault displays, and spare parts management.

Hassia Redatron

"It is generally a contradiction to equip packaging machines so that they combine maximum output with maximum flexibility," notes Andreas Hollmann, Form/Fill/Seal Machine Sales Manager for Hassia Redatron. These requirements led the company, a member of the IWKA Group (now OYSTAR), to the development of a modular approach to designing control software. Specific objectives included the following:

- Develop applications from standardized logic and motion control software modules, assuring consistent, accessible, supportable, and reusable software

- Create a consistent control concept across machinery, based on the existing HMI

- Use standardized programming languages and templates that allow programming assignments to be outsourced

Diagnostics can be performed directly from the controller, with all system data and faults displayed on the machine's HMI. The controller also supports remote diagnostics by factory-trained personnel. Hollman says that their experience to date working from an existing, pretested modular software platform has shown that it is possible to shorten the time required for program startup by 50%.

Conclusion

Packaging machinery builders implementing the modular principles proposed in part 5 of ISA-88 through best practices in the IEC 61131-3 languages have proven that engineering time can be reduced to develop sophisticated, flexible, high-performance capabilities with the documentation and diagnostics needed to achieve maintainability.

ARC Advisory Group has validated the benefits of the emerging trend of software modularity within a standards-based, Make2Pack environment, as described in this chapter, in no fewer than five analyst reports in the past year. A summary of these reports is as follows:

- *Software Management Strategies for Packaging Machine Builders, March 2005.* "Relay ladder logic no longer has the qualifications to serve the needs of the machine builder today, as software that leverages modularity and encapsulation is simply not supportable in this environment. It is the IEC 61131-3 programming languages that are the most viable alternative for OEMs."

- *Will Make2Pack Precipitate a Manufacturing Performance Revolution? June 9, 2005.* "There is also resistance to change among packaging machinery OEMs . . . However, some OEMs have realized design and post-installation services cost reductions using this modular approach . . . Make2Pack is developing a limited set of common control and equipment modules to enable interoperability and reduce the total cost of ownership in a multi-vendor environment typical in packaging operations."

- *Make2Pack Helps OEMs and Integrators Respond to Changing User Requirements, August 25, 2005.* "Packaging OEMs and systems integrators in the forefront of the Make2Pack effort have an opportunity to improve their financial performance by more effectively responding to changing user requirements, while those who lag too far behind the technology adoption curve increase risk to both their business and financial performance."

- *Integral Robotics Raises Agility and Flexibility of Packaging Machinery, September 2005.* "Integration of a robotic manipulator further leverages the concept of modularity by encapsulating this element of the machine as a functional subcomponent that is using a common time base of the overall machine control system . . . The IEC 61131 industrial programming standard, which has become the norm for G3 machines, allows automation suppliers to extend the language for robotics application. Automation suppliers that support a high degree of software modularity are encapsulating robotic functionality as a published library of standardized function blocks."

- *Packaging Machinery Strategies for End Users and Machine Builders, November 2005.* "OEM machine builders must adopt the principles of modular software design based on international standards and guidelines such as the ANSI/ISA-88 batch standard and Make2Pack initiative (ISA-88.05), the OMAC packaging guidelines, PackML state model, IEC 61131-3, PLCopen Motion Control Library, and the evolving PackAL library definitions . . . The modular approach enables software and documentation reusability as well as making it easier to understand the functions . . . Benefits include reduced development cost, reduced delivery time, reduced validation and startup time and cost, and improved post installation service capability and margin."

Procedural Control and Exception Handling

William M. Hawkins
Owner
wmh@iaxs.net
HLQ Ltd.
10300 Colorado Road
Suite 342
Bloomington, MN 55438, USA

Procedural Control

Batch control systems are built around procedural control systems. ISA-88.01 is a standard for batch control systems, so it necessarily describes procedural control systems. At this point, a procedural control system can be described by ISA-88.01 if you strip away the models and terminology that only apply to batch control. The PackML group teaches that some things must be added to a basic procedural control system. Other continuous and discrete systems will have their own set of additions. ISA has chartered ISA106 to standardize these additional sets of rules.

A procedural control system is designed to execute "steps" in a procedure. A batch control system is used to make batches of product using procedural control guided by recipes. One control recipe makes one batch of product. This is enormously useful where one set of controlled equipment can make as many different products as there are recipes that can use that equipment (see ISA-88.01).

Procedural control can be applied to other procedures that do not make products, such as Clean In Place (CIP) and starting up a distillation column. Units are not restricted to producing one batch at a time. Recipes are not necessary unless there is an advantage to using the same procedure with different parameter sets. The ISA-88 recipe management and information management functions would be less complex if products were not made.

Procedural elements remain the same as described in ISA-88.01, although there may be some resistance to the name "Phase" for the smallest element that can provide a process action. Operations still can't be done in parallel in a unit because an operation runs a set of process actions that must have exclusive control of the unit.

Coordination control is required if procedures interact with or cross unit boundaries. Allocation may be necessary, especially for CIP, which forces cleaning fluids through a selection of pipes, valves, and tanks. Part of a matrix of tanks and valves may be handling beer, while another handles cleaning. Mixing beer and solvent is obviously not good for the product.

Characteristics of Continuous Processes

Continuous processes have block valves to isolate various pieces of equipment for maintenance, and they are all manual. Some equipment, like air dryers or switch condensers, may use automatic block valves operated by a simple timed sequencer in order to keep product flowing—not to make batches.

Control valves almost never have limit switches. Valves connected to a digital fieldbus may return accurate position information. A valve is considered closed if a signal of –5% is applied to the pneumatic converter or the positioner.

Control is mostly Proportional Integral Derivative (PID), trying to hold steady process conditions. Refineries are likely to have multiple cascade loops that set a flow loop, such as an optimization program setting an analyzer loop setpoint that sets a temperature loop. Override control using control output selectors may be used to provide constraint control for distillation columns and others. Part of the startup procedure will involve closing cascades and enabling override controls at the proper times as process conditions approach normal.

It may be necessary to change control schemes as the process changes regimes (which are less defined than states) on the way from cold startup to normal operation or back down again. Model-based control is generally restricted to normal operation because the mathematical predictions fall outside of some region near normal. Autotuning may need to be disabled, allowing preset tuning changes as conditions change.

Buffer tanks were used to allow maintenance on a unit process while the upstream line kept running. Improvements in equipment and control have allowed buffer tanks to be reduced or eliminated. Tanks can speed up startups because a quantity of material is available to get a distillation column running on total reflux before the upstream units are running.

Characteristics of Discrete Processes

Discrete unit processes make some physical change in a discrete product, as opposed to the chemical changes or mixing in a continuous process. There are machines that make parts and machines that assemble parts and perhaps paint them. There may be many material transfers to supply raw materials to parts machines, finished parts to assembly machines, and assemblies to final assembly machines. Each of those transfers ends with orienting the part or assembly for use by the next machine. One-armed stationary or mobile robots are useful for transfers.

Statistical quality control is used because it is not economical or necessary to measure every part. A machine whose output fails the sampled tests is stopped so that the problem can be corrected. There is almost no concern for constraints of pressure, temperature, or level. There is great concern for position, so that a part or assembly is aligned properly with the machine that will do something to it. Position may be checked by microswitches or vision systems. A misalignment signal must stop the machine. Drilling a hole in the wrong place ruins the entire part or assembly.

While many product alarms are generated by statistical sampling and alarmed after a delay, some properties may be measured directly, such as connection resistance. The position of a bad part may have to be remembered until it gets to a place where it can be discarded.

Motion control systems have replaced gears and cams in modern systems. The servo motors have to be initialized after being turned on, so that their shafts will be synchronized in position when the machine is started.

Note that there is an infinite variety of discrete manufacturing machines. The list of systems described in this section is far from complete.

Applying Batch Control to Discrete and Continuous Processes

Today, vendors that caught the wave of ISA-88.01 have batch control systems that can only do batch control. Recipes are required, and units must make one part of a batch at a time and give that batch an ID. For an example of the contortions necessary to make CIP fit a batch control system, see chapter 11 of volume 1, *ISA-88 Implementation Experiences*.

Batch control systems are not what continuous and discrete processes need. There are many terms, models, and rules that are completely unfamiliar to those who design, use, and maintain those processes. ISA-88 will not help control system sales beyond batch until vendors stop talking about "batch engines" and "batch managers" and "batch historians" and replace "batch" control with "procedural" control.

Exception Handling

It is not enough to be able to automate the normal execution of a manufacturing procedure. Everyone in manufacturing operations knows that things do not always go as planned. Production equipment and supplies, control equipment, and external utilities may fail. The weather may bring a tornado or a deep freeze.

The supply chain may have a problem that delays "just-in-time" materials. It is even possible that a poorly trained or overstressed operator may do the equivalent of dropping a wrench into a gearbox. These things cause exceptions to the planned execution of a procedure.

All process control systems should have some way to tell an operator that things are not proceeding normally. We have had alarm systems almost as long as we have had automatic process control. One of the vendors in the sixties called this "operation by exception." The operator didn't have to focus on the process for the duration of his shift because an alarm would tell him when an exception from normal control occurred.

Procedural control automates what an operator could do by following a written procedure. It is important to notify an operator when an exception to normal procedural control occurs, but it may be more important to do something now. Several minutes from now may be too late (as in the three stages of tank level—high, too high, and too late). Exception handling is added to procedural control in order to alter the normal procedure quickly so that things don't get worse. For example, a beer tank farm requires that a cross-connection between CIP and product be handled in 8 milliseconds.

The process still requires alarms for violations of physical constraints when no procedures are active to detect and handle exceptions. A leaking valve could fill a tank. A leaking steam valve could overheat something. Dangerous processes may require a separate safety shutdown system that acts when all else fails. Exception handling requires a functional computer and programmers that can think of everything. The shutdown system independently detects constraint violations (e.g., pressure, temperature, level) and acts to abort the process.

All procedural elements should begin by checking that it is safe to proceed and then create an exception if it isn't safe.

State Machines

Whatever executes the automated procedure needs to have a set of states that can change the normal procedure state to one suitable for handling the exception. This produced the now-famous state model shown in figure 18 of ISA-88.01. When PackML showed great interest in that model, those of us that weren't sure the model even belonged in the standard were perplexed. It became apparent that the attraction came from finding a way to unify the way that diverse packaging machines handled exceptions.

Exception handling is simplified by choosing some process states and describing a state machine for them. ISA-88.01 chose Hold, Stop, and Abort. ISA-88.01 section 5.8, "Exception Handling," recognizes three classes of responses to exceptions:

- Temporarily suspend the procedure and stop adding material and energy (Hold)

- Permanently cease the procedure in an orderly manner (Stop)

- Bring the energy stored in the process to zero now, regardless of what might break (Abort or Emergency Stop)

Figure 17.1 shows the ISA-88.01 state machine with a few modifications. The first modification is to add an Initialize state so that multiple procedures may be run in series without going all the way back to Idle. The Initialize state could run applications that get the next procedure and set it up to run. The second modification is to remove the Pause state. It is not significantly different from the Hold state. Alternatively, the operator can set the procedure mode to Manual or Single Step. This stops automatic procedural control and gives responsibility to the operator who saw some need to interrupt automatic operation. The holding procedure does not stop the normal procedure if it is performing steps that must be executed as a complete sequence, until that sequence is complete.

There are times when Abort and Stop are handled the same way, if the process doesn't store enough energy to be dangerous. The exits for Stopped and Aborted require an operator Reset command before going back to Idle because there may be a big mess to clean up. The Reset command for Complete is optional, as is the Start command from Initialize if the procedure normally cycles.

The PackML model replaces Running with two alternating states—Standby and Producing—as shown in Figure 17.2. The state changes from Standby to Producing when machine sensors detect no exceptions and back to Standby when an exception is detected. The exceptions that cause this behavior are usually temporary and self-healing, such as an empty feed bin or the delay while a new pallet is moved into place. If the machine jams in a way that requires human intervention, then Producing would go to Holding and Held.

One way to use the state machine is to include code for Running, Holding, Restarting, Stopping, and Aborting in each master procedure, unit procedure, Equipment Module (EM) procedure, and phase within a procedure. Phases have state detectors and switch to the code for a new state in order to handle the exception, depending on where the phase was when the state changed.

There may be state machines at each of these levels, but not for a phase. Phases do not normally have state machines because phases can run in parallel. If one phase goes into Hold or Stop, it might not make any sense to let the other parallel phases continue. It seems better to have the phase report what it detected to the next level, and let that level decide which state to enter.

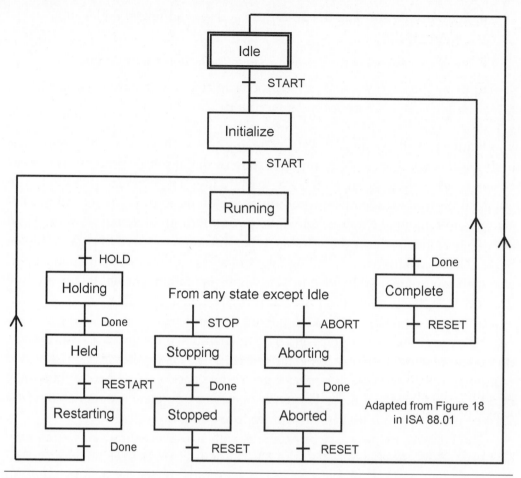

Figure 17.1. State machine for handling procedural control exceptions.

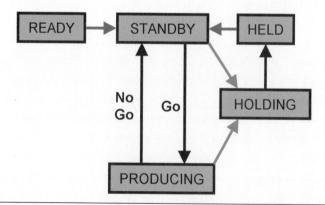

Figure 17.2. Part of the PackML state machine, extracted from Figure 10.4.

Figure 17.3 adds a master control level to the physical model of ISA-88.01, which is useful for machines that are composed of units. Even if the machine has only one unit, master control provides a consistent operator interface for functions that affect an entire machine, such as Start, Hold, Stop, and Reset.

The diagram shows that cell 7 has three machines controlled by master 7.1, 7.2, and 7.3. Each master has three units, and each unit has three EMs. (How many were going to St. Ives?) The numbers may be changed to suit your situation. Naturally, each EM has one or more Control Modules (CMs), at least one of which must have at least one sensor and an actuator that manipulates the process.

CMs don't have state machines because they don't have procedural control (except possibly sequences that are built from basic control elements and that do not change when the EM procedure changes). CMs do exception handling with basic control elements such as interlocks.

Each block in the diagram may have its own state machine, as described previously. Changes in state propagate down the levels because you don't want any modules to continue to run when the master asserts Stop.

Normally, each level is in the Running state, where it executes the normal procedure and detects exceptions. When an exception is detected, the detector changes the state for that level and all levels below it. An upper level may detect that a lower level has changed state and take appropriate action. The state at each level may be changed by an operator command (i.e., Start, Hold, Restart, Stop, Abort, or Reset).

Designing Exception Handlers

The use of state machines to direct normal execution to an exception handling procedure has been described previously. This section is about designing exception handling procedures.

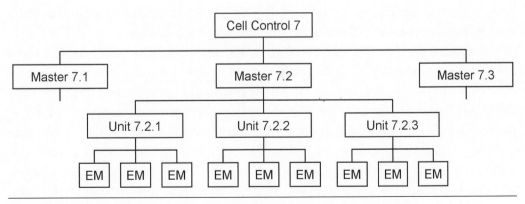

Figure 17.3. Hierarchy of procedural control execution and state machines.

It is said that a man's toolbox only needs two items:

- If it doesn't move when it should, use WD-40
- If it moves when it shouldn't, use duct tape

While this could apply to a process, you could also change "move" to open, full, hot, flow, and so on. These things, however, cannot be fixed by a water displacement spray or heavy-duty adhesive tape. Other kinds of exceptions include the following:

- Something just isn't available, like a material, solvent, utility, equipment, or operator
- Control or equipment failure occurs and requires replacement
- The process comes too close to a physical constraint, which trips an interlock or safety shutdown system
- An external event occurs, like a fire fed by a fuel leak or flooding from a river

The first thing to do is to make an exhaustive list of the exceptions that could occur. Understand what could go wrong with the process and go through manuals for the equipment and systems. Look for product quality exceptions and those caused by waiting too long for something to happen. There are formal procedures for doing this, particularly the hazard and operability study known as HAZOP.

Eliminate the exceptions that are too expensive to fix or have an acceptably low risk. Next pick out those exceptions that are best handled by basic interlocks or a safety shutdown system, which makes them independent of any procedure being followed. Finally, pick out those that can't be handled by procedural control, such as management problems or anything requiring a lawyer.

Consider the possible effects of allowing an operator to issue control commands like changing the mode or setpoint of a PID controller while procedural control is running in an automatic mode. This can add many exceptions to be handled. It is better to prevent operator changes while normal procedural control is in progress. The operator can operate after changing the procedural control mode to Manual, and the operator is responsible for preparing the process to return to Auto procedural control in the Resuming state. The Resuming exception handler will check to see if it is safe to proceed and kick the mode back to Manual if it isn't. It would be good to tell the operator why it went back to Manual.

Detecting and Handling Exceptions

With a list of exceptions to be handled, the next step is to discover how to detect each exception. Most of them will already have alarms from basic control. Some can be detected by a combination of events or a deviation from a profile. Search the Web for "fault signature analysis" to learn about finding faults in complex systems. Timing faults are detected by setting timers or calculating time intervals.

There will be times when a fault is not a fault unless it occurs in a specific part of a procedure, such as a valve position, temperature, pressure, or level. Such faults have to be detected in the normal procedure. This could be done by setting the trip point of an alarm detector, rather than continuous polling. Do not do this with alarms set near physical constraints for pressure, temperature, and level that are connected to interlocks.

Fault handling most often is done by simply giving a Hold command to the appropriate state machine when an alarm is triggered. This can be done with a table of alarms and actions to be taken, so that when an alarm occurs, the table is searched for the appropriate action to take (e.g., sending Hold, Stop, or Abort to a specific state machine).

Where possible, the procedural exception handler should restore normal control and resume. If this is not possible, an operator should be alerted by an alarm that occurs before the condition becomes an exception.

Summary

ISA-88.01 describes good engineering practices for automating procedures. These practices can be applied to any process that can benefit from automated procedures. Exception handling is required for life in a world where apparent certainties are really only probabilities with values less than one. The ISA-88 state machine provides a method for handling failed certainties in a uniform and understandable manner.

Further information can be found on these topics in *Batch Control Systems* by Hawkins and Fisher, published by ISA Press in 2006.

Index

accountability gap, 157–58
auditable systems, 37
automation
    batch techniques for continuous
      process management and control, 65
    continuous processing facilities,
      73–84
    ISA-88.01 equipment state transition
      model, 122–23
    PACKML ISA-88.01 state model,
      108–9
    semi-continuous systems, 87–94
    software modularity, 200–210

batch processing techniques
    automation characteristics, 65
    blended batch-continuous
      technologies, 66
    continuous process management
      and control, 61–71, 74–75, 213
    discrete processes, 213
    material equipment modules, 67–68
    operating characteristics, 64–65
    packaging systems, 139–47
    recipe management, 66
    semi-continuous systems, 86–100
    standards-based control, 33–35
    startup sequencer, 66–67
blister packaging systems, 189–92
buffering units, 19–21
business models
    compliant packaging, 188–89

in Consumer Packaged Goods (CPG)
    operations, 104–6
equipment process statecharts,
    133–34
high-frequency short duration
    stoppages, 154
MCN security, 165–67
original equipment manufacturers,
    168–70

capital investment, reduction of, 5
change control and configuration
  management
    compliant packaging systems,
      194–95
    single control systems, 58
Clean In Place (CIP) execution, 131–32
compliant packaging environment
    business drivers, 188–89
    change management, 194–95
    documentation, 195–96
    governance issues, 198
    ISA-88 standard, 189–92
    manufacturing process, 192–94
    regulatory compliance, 194
    shop floor integration, 189
    training framework, 196–98
condition logic, 25–27
Consumer Packaged Goods (CPG)
  operations
    business drivers in, 102–3
    ISA-88 standard and, 102–4